UNDERSTANDING VIRUSES

UNDERSTANDING VIRUSES

By

Dr. Shubhrata R. Mishra
Deptt. of Botany
Vikram University
Ujjain (M.P.)
(India)

DISCOVERY PUBLISHING HOUSE PVT. LTD.
NEW DELHI-110 002

Published by:
Tilak Wasan
DISCOVERY PUBLISHING HOUSE PVT. LTD.
4831/24, Ansari Road, Prahlad Street
Darya Ganj, New Delhi-110002 (India)
Phone: +91-11-23279245, 43764432
Fax: +91-11-23253475
E-mail: parul.wasan@gmail.com
info@discoverypublishinggroup.com
discoverypublishinghouse@gmail.com
web: www.discoverypublishinggroup.com

First Edition: **2011**
ISBN: 978-81-8356-867-8

Understanding Viruses

Printed in:
INDIA

PREFACE

The present title *"Understanding Viruses"* provides a structured approach to learning by covering all the important topics in a uniform, systematic format. The book has been comprehensively designed incorporating recent advances in this fast moving field. It is written to provide accessible information on viruses in compact form for undergraduate students in biology and related life sciences. It will be useful for both beginning students and those who are more advanced. In addition, busy lecturers who require a quick reference compendium will find it useful, particularly for tutional planning. Simple, yet hopefully clear figures and tables are provided throughout the book.

The over-riding goal of this book, and indeed of the whole *Understanding series,* is to present the essential information concering microbiology in a compact, readily accessible form which leads itself to student learning and revision. The convergence of various approaches has generated a rich panorama of detail, the significance of which we are still attempting to unraval. The present text has been written as an introduction to this rapidly growing field.

To make the work more comprehensive and informative, the author has consulted many authoritative books, research journals, abstracts, monographs etc., so there can be no claim to originality except in the manner of treatment.

The author expresses his thanks to his friends and colleagues whose continue inspirations have initiated him to bring out this book.

The author expresses his gratitude to Mr. Wasan and staff of M/s Discovery Publishing House Pvt. Ltd. for their whole hearted co-operation in the publication of this book.

In the mean time, the author will remain sincerely responsible for any shortcomings of the book and be grateful to the readers for their suggestions and constructive criticism for the continuous betterment of the book. He takes this opportunity to appeal to the readers to send their suggestions straightaway to his Publisher.

Author

CONTENTS

1

INTRODUCTION

VIRUSES ARE UBIQUITOUS ON EARTH

Viruses infect all cellular life forms: eukaryotes (vertebrate animals, invertebrate animals, plants, fungi) and prokaryotes (bacteria and archaea). The viruses that infect prokaryotes are often referred to as bacteriophages, or phages for short.

The presence of viruses is obvious in host organisms showing signs of disease. Many healthy organisms, however, are hosts of non-pathogenic virus infections, some of which are active, while some are quiescent. Furthermore, the genomes of many organisms contain remnants of ancient virus genomes that integrated into their host genomes long ago. As well being present within their hosts, viruses are also found in soil, air and water. Many aqueous environments contain very high concentrations of viruses that infect the organisms that live in those environments.

There is a strong correlation between how intensively a species is studied and the number of viruses found in that species. Our own species is the subject of most attention as we have a vested interest in learning about agents and processes that affect our health. It is not surprising that there are more viruses known that infect mankind than any other species, and new human viruses continue to be found. The intestinal bacterium *Escherichia coli* has also been the subject of much study and many viruses have been found in this species. If other species received the same amount of attention it is likely that many would be found to be hosts to similar numbers of viruses.

It is undoubtedly the case that the viruses that have been discovered represent only a tiny fraction of the viruses on the Earth. Most of the known plants, animals, fungi, bacteria and archaea have yet to be investigated for the presence of viruses, and new potential hosts for viruses continue to be discovered. Furthermore, the analysis of DNA from natural environments points to the existence of many bacterial species that have not yet been isolated in the laboratory; it is likely that these 'non-cultivable bacteria' are also hosts to viruses.

Reasons for Studying Viruses

Some Viruses Cause Disease

Viruses are important agents of many human diseases, ranging from the trivial (e.g. common colds) to the lethal (e.g. rabies), and viruses also play roles in the development of several types of cancer. As well as causing individuals to suffer, virus diseases can also affect the well-being of societies. Smallpox had a great impact in the past and AIDS is having a great impact today.

There is therefore a requirement to understand the nature of viruses, how they replicate and how they cause disease. This knowledge permits the development of effective means for prevention, diagnosis and treatment of virus diseases through the production of vaccines, diagnostic reagents and techniques, and antiviral drugs. These medical applications therefore constitute major aspects of the science of virology.

Veterinary virology and plant virology are also important because of the economic impact of the many viruses that cause disease in domestic animals and crop plants: foot and mouth disease virus and rice yellow mottle virus are just two examples. Another area where viruses can cause economic damage is in the dairy industry, where phages can infect the lactic acid bacteria that are responsible for the fermentations that produce cheese, yogurt and other milk products.

Some Viruses are Useful

Some viruses are studied because they have useful current or potential applications.

1. *Phage typing of bacteria.* Some groups of bacteria, such as some *Salmonella* species, are classified into strains on the basis of the spectrum of phages to which they are susceptible. Identification of the phage types of bacterial isolates can provide useful epidemiological information during outbreaks of disease caused by these bacteria.
2. *Sources of enzymes.* A number of enzymes used in molecular biology are virus enzymes. Examples include reverse transcriptases from retroviruses and RNA polymerases from phages.

3. *Pesticides.* Some insect pests are controlled with baculoviruses and myxoma virus has been used to control rabbits.
4. *Anti-bacterial agents.* In the mid-20th century phages were used to treat some bacterial infections of humans. Interest waned with the discovery of antibiotics, but has been renewed with the emergence of antibiotic-resistant strains of bacteria.
5. *Anti-cancer agents.* Genetically modified strains of viruses, such as herpes simplex virus and vaccinia virus, are being investigated for treatment of cancers. These strains have been modified so that they are able to infect and destroy specific tumour cells, but are unable to infect normal cells.
6. *Gene vectors for protein production.* Viruses such as certain baculoviruses and adenoviruses are used as vectors to take genes into animal cells growing in culture. This technology can be used to insert into cells genes encoding useful proteins, such as vaccine components, and the cells can then be used for mass production of the proteins.
7. *Gene vectors for treatment of genetic diseases.* Children with severe combined immunodeficiency (baby in the bubble syndrome) have been successfully treated using retroviruses as vectors to introduce into their stem cells a non-mutated copy of the mutated gene responsible for the disease.

Virus Studies have Contributed to Knowledge

Much of the basic knowledge of molecular biology, cell biology and cancer has been derived from studies with viruses. Here are a few examples.

1. A famous experiment carried by Alfred Hershey and Martha Chase, and published in 1952, used phage T2 and *E. coli* to provide strong evidence that genes are composed of DNA.
2. The first enhancers to be characterized were in genes of simian virus 40 (SV40).
3. The first transcription factor to be characterized was the transplantation (T) antigen of SV40.
4. The first nuclear localization signal of a protein was identified in the T antigen of SV40.
5. Introns were discovered during studies of adenovirus transcription.
6. The role of the cap structure at the 5' end of eukaryotic messenger RNA was discovered during studies with vaccinia virus and a reovirus.

7. The first internal ribosomal entry site to be discovered was found in the RNA of poliovirus.
8. The first RNA pseudoknot to be discovered was that in the genome of turnip yellow mosaic virus.

Nature of Viruses

Viruses are Small Particles

Evidence for the existence of very small infectious agents was first provided in the late 19th century by two scientists working independently: Martinus Beijerinck in Holland and Dimitri Ivanovski in Russia. They made extracts from diseased plants, which we now know were infected with tobacco mosaic virus, and passed the extracts through fine filters. The filtrates contained an agent that was able to infect new plants, but no bacteria could be cultured from the filtrates. The agent remained infective through several transfers to new plants, eliminating the possibility of a toxin. Beijerinck called the agent a 'virus' and the term has been in use ever since.

At around the same time, Freidrich L¨oeffler and Paul Frosch transmitted foot and mouth disease from animal to animal in inoculum that had been highly diluted. A few years later Walter Reed and James Carroll demonstrated that the causative agent of yellow fever is a filterable agent. Figure gives some indication of the size of these agents, which are known as virus particles or virions. The virion of a herpesvirus, which is a fairly large virus, is about ten million times smaller than a large balloon, while the balloon is smaller than the Earth by the same factor. The virions of most viruses are too small to be seen with a light microscope and can be seen only with an electron microscope. The units in which virions are normally measured are nanometres (1 nm = 10^{-9} m). Although virions are very small, their dimensions cover a large range. Amongst the smallest are parvoviruses, with diameters about 20 nm, while the microbe-mimicking virus (mimivirus), isolated from an amoeba, is amongst the largest.

Virology is therefore concerned with very small particles, though often with very large numbers of those particles! A concentrated suspension of virions might contain 10^{12} virions/ml. A single virus-infected cell might produce 10^{5} virions. A person infected with HIV might produce 10^{11} virions in a day.

Virions are not cells. They do not contain organelles, except for the virions of the arenaviruses, which contain cell ribosomes that were packaged when the virions were assembled.

Viruses Have Genes

The virion contains the genome of the virus. Whereas the genomes of cells are composed of double- stranded DNA, there are four possibilities for a virus genome:

1. double-stranded DNA
2. single-stranded DNA
3. double-stranded RNA
4. single-stranded RNA.

The genome is enclosed in a protein coat known as a capsid. The genome plus the capsid, plus other components in many cases, constitute the virion. The functions of the virion are to protect the genome and to deliver it into a cell in which it can replicate.

Generally, virus genomes are much smaller than cell genomes and the question arises as to how viruses encode all their requirements in a small genome. Viruses achieve this in a number of ways.

1. *Viruses use host cell proteins.* The genomes of large viruses duplicate some of the functions of the host cell, but the small viruses rely very heavily on functions of the host cell. There is, however, one function that an RNA virus must encode, no matter how small its genome. That function is an RNA polymerase, because cells do not encode enzymes that can replicate virus RNA. A significant proportion of the genome of an RNA virus is taken up with the gene for an RNA polymerase.
2. *Viruses code efficiently.* There may be overlapping genes and genes encoded within genes. The small genome of hepatitis B virus is a good example.
3. *Many virus proteins are multifunctional.* A virus protein may have several enzyme activities.

Viruses are Parasites

Viruses differ from cells in the way in which they multiply. A new cell is always formed directly from a pre-existing cell, but a new virion is never formed directly from a pre-existing virion. New virions are formed by a process of replication, which takes place inside a host cell and involves the synthesis of components followed by their assembly into virions. Viruses are therefore parasites of cells, and are dependent on their hosts for most of their requirements, including:

1. building-blocks such as amino acids and nucleosides;
2. protein-synthesizing machinery (ribosomes);
3. energy, in the form of adenosine triphosphate.

A virus modifies the intracellular environment of its host in order to enhance the efficiency of the replication process. Modifications might include production of new membranous structures, reduced expression of cell genes or enhancement of a cell process. Some large phages encode proteins that boost photosynthesis in the cells of their photosynthetic bacterial hosts, thereby probably boosting the yields of virus from the cells.

Are viruses living or nonliving?

There is an ongoing debate as to whether viruses are living or nonliving; the view taken depends on how life is defined. Viruses have genes and when they infect cells these genes are replicated, so in this sense viruses are living. They are, however, very different to cellular life forms, so Dr. McCoy's stock phrase on finding new life forms in the galaxy could be applied to viruses. When viruses are outside their host cells they exist as virus particles (virions), which are inert, and could be described as nonliving, but viable bacterial spores are inert and are not considered to be nonliving.

When Beijerinck selected the word 'virus' he chose the Latin word for poison. This term has now been in use for over a century and virology has developed into a huge subject. More recently, the term virus has acquired further meanings. Computers are threatened by *infection* with viruses that can be found in the wild once they have been *released* by their authors. These viruses are *specific* for certain file types. *Infected* files may be put on several web sites and a virus *epidemic* may ensue. Another use of the term virus is exemplified in John Humphrys' book 'Lost For Words', in which he talks about the *deadly* virus of management-speak *infecting* language.

Having outlined the nature of viruses and why they are important, the remainder of the book will examine many aspects of fundamental

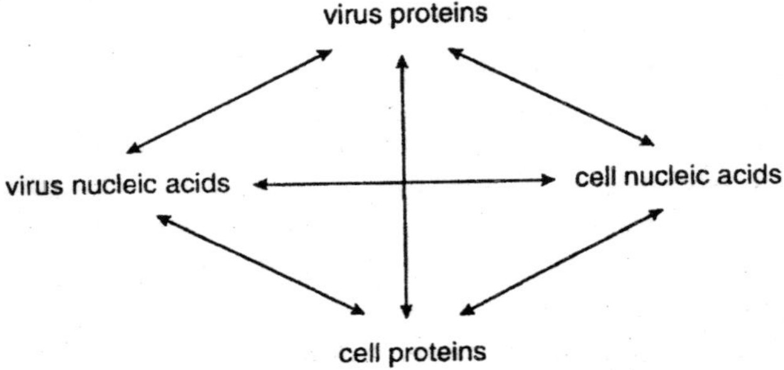

Fig. 1.1. Interactions between virus molecules and cell molecules.

and applied virology. The early chapters cover principles, such as the structure of virions, virus replication and the classification of viruses. There are then nine chapters devoted to reviews of particular groups of viruses, where both principles and applications of virology are covered. Towards the end of the book we consider specific applications of virology, including viral vaccines and anti-viral drugs. The final chapter is devoted to prions, which are not viruses! It is important to point out that much of virology is concerned with characteristics of the proteins and nucleic acids of viruses, and with interactions between these molecules and the proteins and nucleic acids of cells. Most of these interactions rely on specific binding between the molecules.

2

Viruses in Lab

Methods used in virology are introduced early in this book in order to provide an appreciation of the nature of the techniques that have been used to achieve our current level of knowledge and understanding of viruses. Virology is a huge subject and uses a wide range of methods. Many of the techniques of molecular biology and cell biology are used, and constraints on space permit us to mention only some of them. Much of the focus of this chapter is on methods that are unique to virology. Many of these methods are used, not only in virus research, but also in the diagnosis of virus diseases of humans, animals and plants.

Cultivation of Viruses

Virologists need to be able to produce the objects of their study, so a wide range of procedures has been developed for cultivating viruses. Virus cultivation is also referred to as propagation or growth, all terms borrowed from horticulture! A few techniques have been developed for the cultivation of viruses in cell-free systems, but in the vast majority of cases it is necessary to supply the virus with appropriate cells in which it can replicate.

Phages are supplied with bacterial cultures, plant viruses may be supplied with specially cultivated plants or with cultures of protoplasts (plant cells from which the cell wall has been removed), while animal viruses may be supplied with whole organisms, such as mice, eggs containing chick embryos or insect larvae. For the most part, however, animal viruses are grown in cultured animal cells.

Animal Cell Culture

Animal cell culture techniques are well developed and most of the cells used are from continuous cell lines derived from humans and

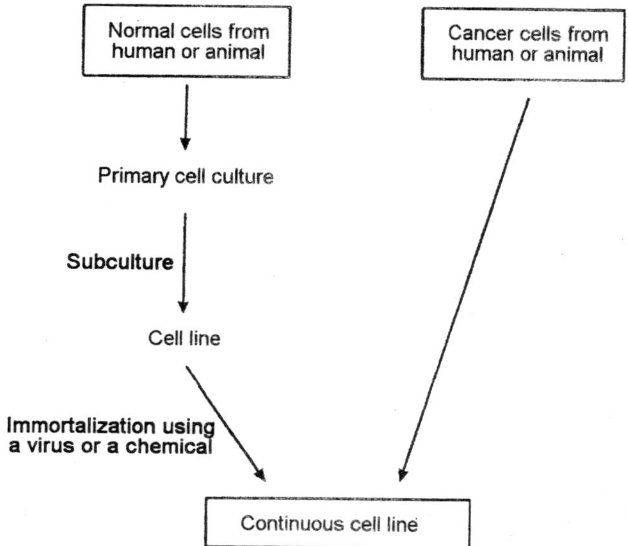

Fig. 2.1. Derivation of continuous cell lines of human and animal cells.

other animal species. Continuous cell lines consist of cells that have been immortalized, either in the laboratory or in the body; they can be subcultured indefinitely. The HeLa cell line is a widely used continuous cell line that was initiated in the middle of the 20th century from cells taken from a cervical carcinoma.

Sometimes it is difficult to find a cell line in which a virus can replicate. For many years no suitable cell culture system could be found for hepatitis C virus, but eventually a human hepatoma cell line was found to support replication of an isolate of the virus.

Cells are cultured in media that provide nutrients. Most media are supplemented with animal serum, which contains substances that promote the growth of many cell lines. Other important roles for the medium are the maintenance of optimum osmotic pressure and pH for the cells. Viruses can be cultivated in cells growing on the surface of a variety of plastic vessels with the cells bathed in the growth medium. Most cells grow on a plastic or glass surface as a single layer of cells, known as a monolayer. Alternatively the cells can be suspended in the medium, which is stirred to keep them in suspension.

Contamination with bacteria and fungi can cause major problems in cell culture work; in order to minimize these problems work is normally done in a sterile cabinet and most media contain antibiotics.

Many cell types require a relatively high concentration of carbon dioxide, which can be supplied in a special incubator.

Isolation of Viruses

Many viruses can be isolated as a result of their ability to form discrete visible zones (plaques) in layers of host cells. If a confluent layer of cells is inoculated with virus at a concentration so that only a small proportion of the cells is infected, then plaques may form where areas of cells are killed or altered by the virus infection. Each plaque is formed when infection spreads radially from an infected cell to surrounding cells.

Plaques can be formed by many animal viruses in monolayers if the cells are overlaid with agarose gel to maintain the progeny virus in a discrete zone. Plaques can also be formed by phages in lawns of bacterial growth.

It is generally assumed that a plaque is the result of the infection of a cell by a single virion. If this is the case then all virus produced from virus in the plaque should be a clone, in other words it should be genetically identical. This clone can be referred to as an isolate, and if it is distinct from all other isolates it can be referred to as a strain. This is analogous to the derivation of a bacterial strain from a colony on an agar plate.

There is a possibility that a plaque might be derived from two or more virions so, to increase the probability that a genetically pure strain of virus has been obtained, material from a plaque can be inoculated onto further monolayers and virus can be derived from an individual plaque. The virus is said to have been plaque purified.

When a virus is first isolated it may replicate poorly in cells in the laboratory, but after it has gone through a number of replication cycles it may replicate more efficiently. Each time the virus is 'sub-cultured' (to borrow a term from bacteriology) it is said to have been passaged. After a number of passages the virus may be genetically different to the original wild strain, in which case it is now a laboratory strain.

Centrifugation

After a virus has been propagated it is usually necessary to remove host cell debris and other contaminants before the virus particles can be used for laboratory studies, for incorporation into a vaccine, or for some other purpose. Many virus purification procedures involve centrifugation; partial purification can be achieved by differential

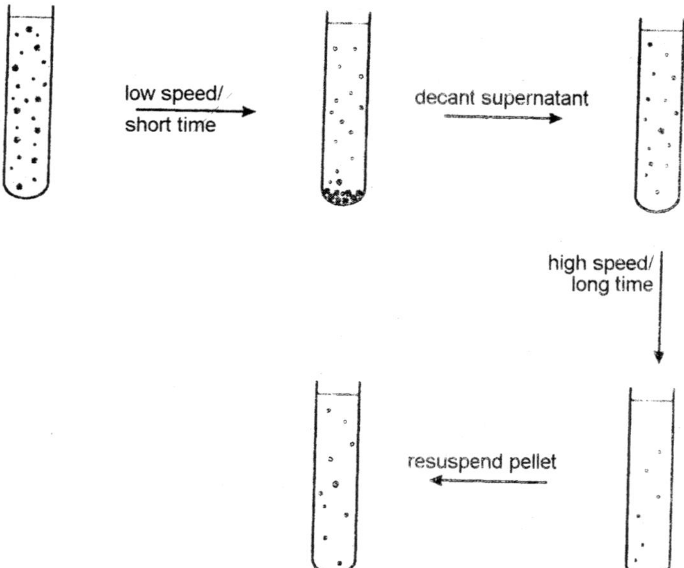

Fig. 2.2. Partial purification of virions by differential centrifugation

centrifugation and a higher degree of purity can be achieved by some form of density gradient centrifugation.

Differential Centrifugation

Differential centrifugation involves alternating cycles of low-speed centrifugation, after which most of the virus is still in the supernatant, and high-speed centrifugation, after which the virus is in the pellet.

Density Gradient Centrifugation

Density gradient centrifugation involves centrifuging particles (such as virions) or molecules (such as nucleic acids) in a solution of increasing concentration, and therefore density. The solutes used have high solubility: sucrose is commonly used. There are two major categories of density gradient centrifugation: rate zonal and equilibrium (isopycnic) centrifugation.

In rate zonal centrifugation a particle moves through the gradient at a rate determined by its sedimentation coefficient, a value that depends principally on its size. Homogeneous particles, such as identical virions, should move as a sharp band that can be harvested after the band has moved part way through the gradient.

In equilibrium centrifugation a concentration of solute is selected to ensure that the density at the bottom of the gradient is greater than

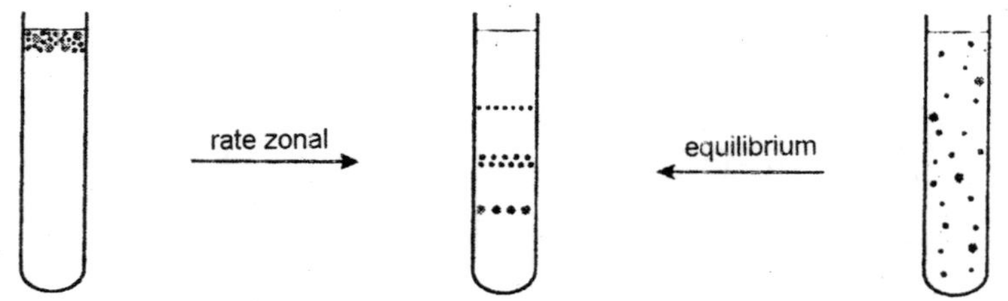

Fig. 2.3. Purification of virions by density gradient centrifugation.

that of the particles/molecules to be purified. A particle/molecule suspended in the gradient moves to a point where the gradient density is the same as its own density. This technique enables the determination of the buoyant densities of nucleic acids and of virions. Buoyant densities of virions determined in gradients of caesium chloride are used as criteria in the characterization of viruses.

Structural Investigations of Cells and Virions

Light Microscopy

The sizes of most virions are beyond the limits of resolution of light microscopes, but light microscopy has useful applications in detecting virus-infected cells, for example by observing cytopathic effects or by detecting a fluorescent dye linked to antibody molecules that have bound to a virus antigen (fluorescence microscopy).

Confocal microscopy is proving to be especially valuable in virology. The principle of this technique is the use of a pinhole to exclude light from out-of-focus regions of the specimen. Most confocal microscopes scan the specimen with a laser, producing exceptionally clear images of thick specimens and of fluorescing specimens. Furthermore, 'optical slices' of a specimen can be collected and used to create a three-dimensional representation. The techniques can be used with live cells and can be applied to investigations of protein trafficking, with the virus or cell protein under investigation carrying a suitable label, e.g. green fluorescent protein (a jellyfish protein).

Electron Microscopy

Many investigations of the structure of virions or of virus-infected cells involve electron microscopy. Large magnifications are achievable with a transmission electron microscope but the specimen, whether it is a suspension of virions or an ultrathin section of a virus-infected cell, must be treated so that details can be visualized.

Negative staining techniques generate contrast by using heavy-metal-containing compounds, such as potassium phosphotungstate and ammonium molybdate. In electron micrographs of virions the stains appear as dark areas around the virions, allowing the overall virion shape and size to be determined. Further structural detail may be apparent if the stains penetrate any crevices on the virion surface or any hollows within the virion. Negative staining techniques have generated many high quality electron micrographs, but the techniques have limitations, including structural distortions resulting from drying.

Cryo-electron microscopy techniques are more recent. In these a wet specimen is rapidly cooled to a temperature below −160°C, freezing the water as a glass-like material. The images are recorded while the specimen is frozen. They require computer processing in order to extract maximum detail, and data from multiple images are processed to reconstruct three-dimensional images of virus particles. This may involve averaging many identical copies or combining images into three-dimensional density maps (tomograms).

X-ray Crystallography

X-ray crystallography is another technique that is revealing detailed information about the three-dimensional structures of virions (and DNA, proteins and DNA–protein complexes). This technique requires the production of a crystal of the virions or molecules under study. The crystal is placed in a beam of X- rays, which are diffracted by repeating arrangements of molecules/atoms in the crystal. Analysis of the diffraction pattern allows the relative positions of these molecules/ atoms to be determined.

Other techniques that are providing useful information about the structure of viruses are nuclear magnetic resonance and atomic force microscopy.

Electrophoretic Techniques

Mixtures of proteins or nucleic acids can be separated by electrophoresis in a gel composed of agarose or polyacrylamide. In most electrophoretic techniques each protein or nucleic acid forms a band in the gel. Electrophoresis can be performed under conditions where the rate of movement through the gel depends on molecular weight. The molecular weights of the protein or nucleic acid molecules can be estimated by comparing the positions of the bands with positions of bands formed by molecules of known molecular weight electrophoresed in the same gel. The technique for estimating molecular weights of proteins is polyacrylamide gel electrophoresis in the presence of the detergent sodium dodecyl sulphate (SDS-PAGE).

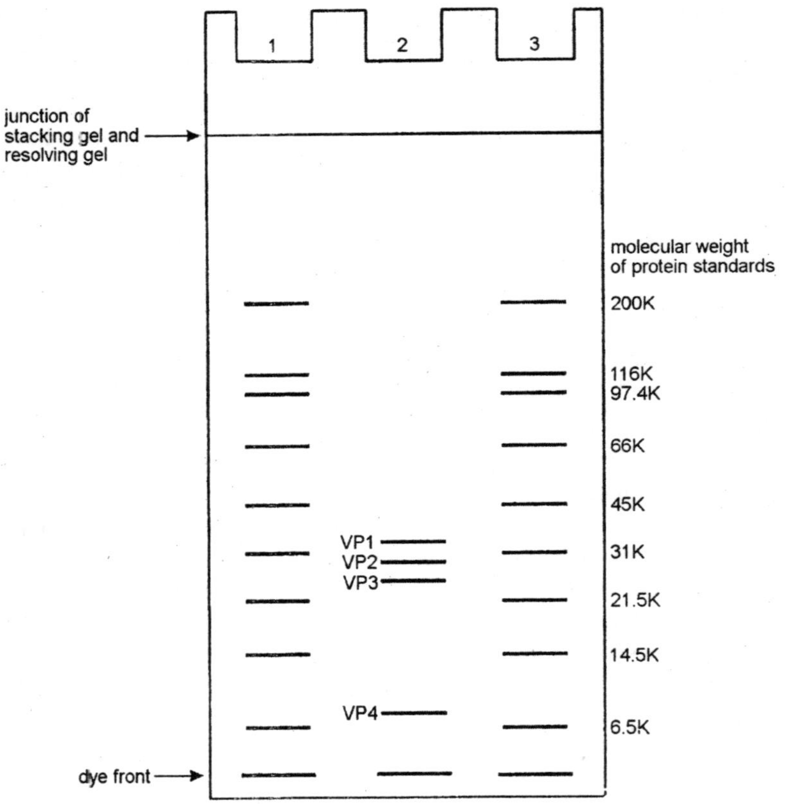

Fig. 2.4. Separation of proteins and estimation of their molecular weights using SDS-PAGE.

The patterns of nucleic acids and proteins after electrophoretic separation may be immobilized by transfer (blotting) onto a membrane. If the molecules are DNA the technique is known as *Southern blotting*, named after Edwin Southern; if the molecules are RNA the technique is known as *northern blotting*, and if the molecules are protein the technique is known as *western blotting*!

Detection of Viruses and Virus Components

A wide range of techniques has been developed for the detection of viruses and virus components and many of them are used in laboratories involved with diagnosis of virus diseases. The techniques can be arranged in four categories: detection of (1) virions, (2) virus infectivity, (3) virus antigens and (4) virus nucleic acids.

Detection of Virions

Specimens can be negatively stained and examined in an electron microscope for the presence of virions. Limitations to this approach are the high costs of the equipment and limited sensitivity; the minimum detectable concentration of virions is about 10^6/ml. An example of an application of this technique is the examination of faeces from a patient with gastroenteritis for the presence of rotavirus particles.

Detection of Infectivity Using Cell Cultures

Not all virions have the ability to replicate in host cells. Those virions that do have this ability are said to be 'infective', and the term 'infectivity' is used to denote the capacity of a virus to replicate. Virions may be non-infective because they lack part of the genome or because they have been damaged.

To determine whether a sample or a specimen contains infective virus it can be inoculated into a culture of cells, or a host organism, known to support the replication of the virus suspected of being present. After incubation of an inoculated cell culture at an appropriate temperature it can be examined by light microscopy for characteristic changes in the appearance of the cells resulting from virus-induced damage. A change of this type is known as a *cytopathic effect* (CPE); examples of CPEs induced by poliovirus and herpes simplex virus. The poliovirus-infected cells have shrunk and become rounded, while a multi- nucleated giant cell known as a *syncytium* (plural *syncytia*) has been formed by the fusion of membranes of herpes simplex virus-infected cells.

Detection of Virus Antigens

Virus antigens can be detected using virus-specific antisera or monoclonal antibodies. In most techniques positive results are indicated by detecting the presence of a label, which may be attached either to the antivirus antibody (direct tests) or to a second antibody (indirect tests). The anti-virus antibody is produced by injecting virus antigen into one animal species and the second antibody is produced by injecting immunoglobulin from the first animal species into a second animal species. Antibodies can have many types of label attached and the labels can be detected using a range of methods.

Detection of Virus Nucleic Acids

Hybridization

Virus genomes or virus messenger RNAs (mRNAs) may be detected using sequence-specific DNA probes carrying appropriate labels. Some of the labels that are used for antibody detection can be used to label the probes.

Table 2.1. Molecules used to label antibodies (and nucleic acids) and techniques used to detect them

Label	*Detection technique*
Enzyme	Enzyme-linked immunosorbent assay
Fluorescent	Fluorescence microscopy
Fluorimetry	
Gold	Electron microscopy
Radio-active	Autoradiography

Hybridization may take place on the surface of a membrane after Southern blotting (DNA) or northern blotting (RNA). Thin sections of tissue may be probed for the presence of specific nucleic acids, in which case the technique is known as *in situ* hybridization.

Polymerase chain reaction (PCR)

When a sample is likely to contain a low number of copies of a virus nucleic acid the probability of detection can be increased by amplifying virus DNA using a PCR, while RNA can be copied to DNA and amplified using a RT (*reverse transcriptase*)-PCR. The procedures require oligonucleotide primers specific to viral sequences. An amplified product can be detected by electrophoresis in an agarose gel, followed by transfer to a nitrocellulose membrane, which is incubated with a labelled probe.

There are also PCR techniques available for determining the number of copies of a specific nucleic acid in a sample. Real-time PCR is commonly used for this purpose. In this technique the increase in DNA concentration during the PCR is monitored using fluorescent labels; the larger the initial copy number of DNA, the sooner a significant increase in fluorescence is observed. The PCR cycle at which the fluorescent signal passes a defined threshold is determined and this gives an estimate of the starting copy number.

Infectivity Assays

An infectivity assay measures the titre (the concentration) of infective virus in a specimen or a preparation. Samples are inoculated into suitable hosts, in which a response can be observed if infective virus is present. Suitable hosts might be animals, plants or cultures of bacterial, plant or animal cells. Infectivity assays fall into two classes: quantitative and quantal.

Quantitative Assays

Quantitative assays are those in which each host response can be any one of a series of values, such as a number of plaques. A plaque assay can be carried out with any virus that can form plaques, giving an estimate of the concentration of infective virus in plaque-forming units (pfu).

The assay is done using cells grown in Petri dishes, or other appropriate containers, under conditions that will allow the formation of plaques. Cells are inoculated with standard volumes of virus dilutions. After incubation, dishes that received high dilutions of virus may have no plaques or very few plaques, while dishes that received low dilutions of virus may have very large numbers of plaques or all the cells may have lysed. Dishes are selected that have plaque numbers within a certain range, e.g. 30–300. The plaques are counted and the concentration of virus in the sample (pfu/ml) is calculated.

Quantal Assays

In a quantal assay each inoculated subject either responds or it does not; for example, an inoculated cell culture either develops a CPE or it does not; an inoculated animal either dies or it remains healthy. The aim of the assay is to find the virus dose that produces a response in 50 per cent of inoculated subjects. In cell culture assays this dose is known as the $TCID_{50}$ (the dose that infects 50 per cent of inoculated tissue cultures). In animal assays the dose is known as the ID_{50}, and where the virus infection kills the animal the dose is known as the LD_{50} (the dose that is lethal for 50 per cent of inoculated animals).

The outcome of a $TCID_{50}$ determination can be used to estimate a virus titre in pfu, or vice versa, using the formula

$$1\ TCID_{50} = 0.7\ pfu$$

This formula applies only if the method used to visualize plaques does not alter the yield of virus.

It is clear that the dilution of virus that contains one $TCID_{50}$ lies between 10^{-3} and 10^{-4}. Several methods have been developed for estimating this end point.

A similar approach can be used to assay the infectivity of any virus that causes the formation of discrete lesions. A number of animal viruses, including smallpox virus and other poxviruses, can cause the formation of pocks after inoculation onto the chorioallantoic membrane of a chick embryo. Similarly, many plant viruses cause the formation of lesions after they have been rubbed on the surface of leaves, though the titration is not directly equivalent to a plaque assay as the number of sites on a leaf through which a virus can enter is limited.

One-step Growth Curve

A common type of experiment that involves assaying virus infectivity is a single-burst experiment, which provides data for a one-step growth curve. The first one-step growth curves produced were for phages, but they have since been produced for other viruses.

For a period (the eclipse period), the titre of infective virus remains constant. During the eclipse period the virus is replicating, but no virus has been released from infected cells, so each infected cell gives rise to one plaque. As infected cells begin to die and release virus the titre begins to rise, and when all the infected cells have lysed the titre levels off.

The procedure for deriving a one-step growth curve is now outlined.

1. A suspension of virions is mixed with a suspension of host cells. It is necessary to ensure that, as far as possible, all the cells are infected simultaneously, so the number of virions must greatly exceed the number of cells. The ratio of virions to cells is known as the multiplicity of infection (m.o.i.). In a one-step growth experiment an m.o.i. between 5 and 10 is commonly used.
2. Adsorption of virions to cells is allowed to proceed for a suitable period (e.g. 2 minutes), and then adsorption is stopped by, for example, greatly diluting the mixture and/or by adding antivirus antiserum.
3. A sample is taken from the suspension at intervals until lysis is complete. For phages the experiment is usually complete in less

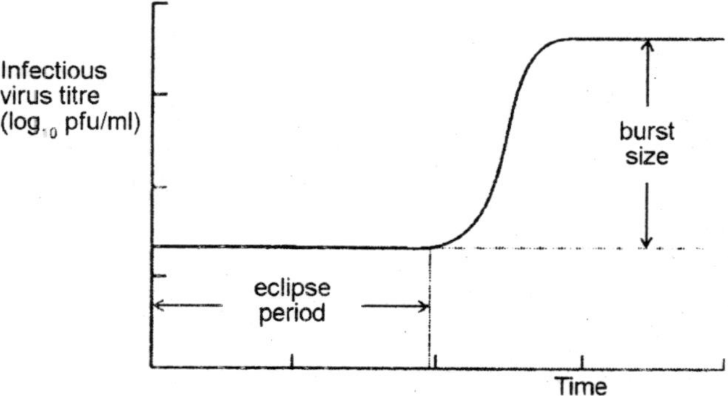

Fig. 2.5. One-step growth curve.

than one hour. For animal viruses, however, the timescale is measured in hours or days, reflecting the much slower growth rates of eukaryotic cells and the longer times taken for replication of animal viruses.

4. A plaque assay is carried out on each sample.
5. The logarithm of the virus titre is plotted against time to obtain the one-step growth curve.

This type of experiment provides valuable information about the replication cycle of a virus in a particular host cell system. The average yield of infectious virus per cell (the burst size) can be calculated from the formula

$$\text{burst size} = \frac{\text{final virus titre}}{\text{initial virus titre}}$$

It should be noted that the burst size is an average of what may be a significant variable as there is usually much variation in burst size between cells. One factor that may affect the burst size is the physiological state of the cells. In one study with a phage, a burst size of 170 was obtained when growing bacteria were used, while a value of 20 was obtained with resting bacteria.

Virus Genetics

Genome Sequencing

Determination of the sequence of bases in a DNA molecule is usually done using the dideoxy chain termination method developed by Fred Sanger and his colleagues at Cambridge. The sequence of the genome of phage φX174 DNA was determined by this group six years after the first determination of a DNA sequence, which itself was from a virus: 12 base pairs (bp) of phage lambda DNA.

With the help of computer programmes such as Artemis and BLAST much useful information can be derived from virus genome sequences.

1. Open reading frames (ORFs) can be found. The ORFs can be translated into the amino acid sequences of the virus proteins, and these may allow characteristics and functions of the proteins to be deduced. Hydrophobic sequences may indicate membrane association; the sequence Gly–Asp–Asp may indicate RNA-dependent RNA polymerase activity.
2. Sequences that regulate gene expression, e.g. promoters and enhancers, can be identified.
3. During an outbreak of virus disease, such as foot and mouth disease or influenza, sequences of virus isolates can be compared. This

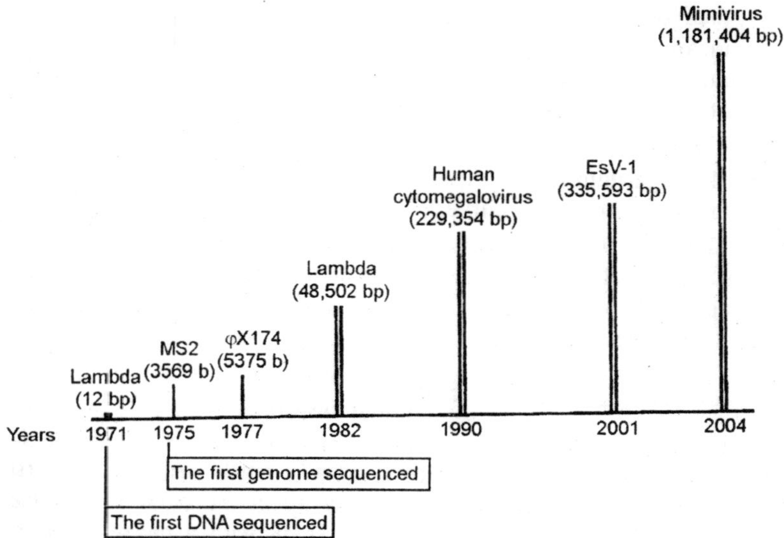

Fig. 2.6. Time line depicting some landmarks in viral genome sequencing.

can provide useful epidemiological information such as the source of the virus responsible and whether more than one strain is responsible for the outbreak.

Genome Manipulation

The wide range of techniques available for the manipulation of nucleic acids can be applied to virus genomes. These techniques include the isolation of specific fragments of genomes using restriction endonucleases, the cloning of fragments in bacterial plasmids and the introduction of site-specific mutations into virus genomes. The natural processes of recombination and reassortment that produce new combinations of virus genes can be harnessed to produce new viral genotypes in the laboratory.

Investigation of Gene Function and Expression

The function of a gene may be deduced if its expression is blocked, and many studies of gene function involve the creation of a virus with a mutated gene. Many of the mutants are unable to replicate under conditions in which the wild-type virus replicates and are known as conditional lethal mutants. Most studies have used temperature-sensitive mutants, which are unable to replicate at temperatures at which the wild-type virus replicates (non-permissive temperatures), but are able to replicate at other temperatures (permissive temperatures).

Techniques for introducing mutations into DNA are well established and have been applied to the study of DNA viruses for some time. More recently, techniques for manipulating the genomes of RNA viruses have been developed. These techniques are referred to as reverse genetics, and they involve the reverse transcription of the RNA genome to DNA, mutation of the DNA, then transcription back to RNA.

Another way in which the expression of a gene can be blocked is by RNA silencing, or RNA interference (RNAi). This approach exploits a cell defence mechanism that can destroy virus mRNAs after induction by specific dsRNA molecules. Short sequences of dsRNA can be used to inhibit the expression of virus genes, and hence to investigate the functions of those genes. Yet another approach to the investigation of gene function involves the introduction of the gene into cells in which it is transiently expressed and changes in the cells are monitored.

As well as the production of virus mutants there are other ways in which viruses can be genetically modified in order to aid their study. A sequence encoding a tag, such as green fluorescent protein, can be appended to the gene for a virus protein so that the distribution of the protein in infected cells can be monitored.

The expression of virus genes in infected cells can be monitored with the aid of DNA microarrays. A DNA microarray consists of a substrate, such as a glass slide, with hundreds, thousands or even millions of small spots of DNA attached. Each spot contains a specific DNA probe and specific DNA or RNA molecules in a sample are detected by hybridization to specific spots.

Microarray technology allows investigators to monitor the expression of hundreds or thousands of genes, and is thus ideal for studies of viruses with large genomes such as the herpesviruses and the large DNA phages. Virus mRNAs are detected by copying RNA from infected cells to DNA using a reverse transcriptase, amplifying the DNA with a PCR, labelling it with a fluorescent dye and adding it to a microarray. The probes that bind DNA from the sample are detected by scanning with a laser at a wavelength that excites the fluorescent dye.

DNA microarrays can also be used to investigate the expression of host cell genes that are relevant to viruses. Cell surface molecules used by viruses as receptors have been identified using microarrays.

Handling Animal Viruses

Viruses are too small to be seen except by electron microscopy and this requires concentrations in excess of 100 particles, or even higher if a virus has no distinctive morphology. Therefore viruses are usually detected by other indirect methods.

These fall into three categories: (1) *multiplication* in a suitable culture system ad detection of the virus by the effects it causes, (2) *serology* which makes use of the inter action between a virus and antibody directed specifically against it, and (3) *nuclei acid probes* which detect unique viral sequences by hybridization.

Selection of Culture Systems

The culture system always consists of living cells and the choice is outlined in Table 2.2 Which is used depends on the aims of the experiment. These may be divided into isolation of viruses, biochemistry of multiplication, structural studies and study of natural infections. The investigation of any new virus starts with ways to cultivate it. There are still many, which are uncultivable, particularly those occurring in the gut, but these can be present in such numbers that they were actually discovered by electron microscopy. Often a virus is suspected of causing a disease. The obvious choice of experimental animal is the natural host, although for man this is ruled out by ethical considerations. Consequently organ cultures and cells must be used. Logically these should be from the natural host and obtained from those sites where the virus multiplies in the whole animal. However, it may prove that cells from unrelated animals are susceptible, e.g. human influenza viruses were first cultivated by inoculating as ferret intranasally and now grow best in embryonated chicken eggs! Frequently viruses grow poorly on initial isolation but adapt on being passed from culture to culture.

Biochemical studies of virus infections require a cell system in which nearly every cell is infected. To achieve this, large numbers of infectious particles, and hence a system which will produce the, is required. Often cells which are suitable for production of virus are different from those used for the study of virus multiplication. There is little logic in choosing a cell system, only pragmatism. Cells differ greatly and different properties make one cell the choice for a particular study and unsuitable for another. Exact control of the cell's environment is needed especially for labeling with radio-isotopes since a chemically defined medium must be prepared that lacks the non-radioactive isotope. Otherwise the specific activity of the radio-isotope would be reduced to an unusable level.

The investigation of natural infections can only be done in the natural host. However these are frequently unsuitable and the nearest

Table 2.2. How to choose a culture system for animal viruses

Culture system	*Advantages*	*Disadvantages*
Animal	Natural infection	Cost of upkeep is expensive. Large variation between individuals even if inbred. Therefore large numbers needed.
Organ, e.g., pieces of brain, gut, trachea	Natural infection	Many cell types present
	Fewer animals needed	Unnatural since isolated from regulating substances such as hormones
	Less variation since one animal gives many organ cultures	
Cell	Can be cloned, therefore variation between individuals in minimal Best for biochemical studies as the environment can be controlled exactly and quickly may be immortal	Very unnatural since cells dedifferentiate when cultured

approximation is usually to use purpose-bred animals, which, although usually not the natural host species, have a similar range of defence mechanisms and can be maintained in the laboratory. The mouse has been extensively studied and inbred strains reduce genetic variability. Although the use of animals for studying virus diseases has been criticized by organizations concerned with animal right, the student of virology will be aware that there is as yet no alternative for studying the complex interactions of viruses with the responses of the host. Analysis of the processes involved would be immensely easier if there were a test-tube system but it is unlikely that any will appear in the foreseeable future except for very specialized purposes.

Organ cultures

These have the advantage of maintaining the differentiated state of the cell. However, there are technical difficulties in their large scale use, and as a result they have not been widely employed. We shall consider here only organ cultures from the trachea. Appropriately, these were first used by an ear, nose and throat surgeon. Bertil Hoorn, who was interested in respiratory viruses. Ciliated cells lining the trachea continue to beat in coordinated waves while the tissue remains healthy. Virus multiplication causes the synchrony to be lost and eventually for the ciliated cells to detach. Virus is also released into fluids surrounding the tissue and can be measured if appropriate assays are available.

Cell cultures

Cells in culture are kept in an isotonic solution consisting of a mixture of salts in their normal physiological proportions supplemented with serum (usually 5-10%) and in such a growth medium cells rapidly adhere to the surface of suitable glass or plastic vessels. Serum is a complex mixture of proteins and other compounds without which mitosis does not occur and only recently have synthetic substitutes become available. Naturally all components used in cell culture are sterile and handled under aseptic conditions to prevent the invasion of micro-organisms. Antibiotics have been invaluable in establishing cells in culture and routine cell culture dates from the 1950s when they first appeared on the market.

Cultured cells are either diploid or heteroploid, having more than the diploid number of chromosomes but not a simple multiple of it. Diploid cells will undergo a finite number of divisions, from around 10-100, whereas the heteroploid cells will divide forever. The latter are known as *continuous cell lines* and they originate from naturally occurring tumours or from some spontaneous event, which alters the control of division of a diploid cell. Diploid cell lines are most easily obtained from embryos by reducing kidneys or the whole body to a suspension of single cells.

Modern methods of cell culture

The methodology described above is suited for research and clinical or diagnostic laboratories but is difficult to scale up for commercial purposes such as vaccine manufacture. There are now various solutions to the problem all aimed at increasing cell density. One of the earliest was to grow cells in suspension and this has been a refined using hybridoma cell for monoclonal antibody production. However many cells only grow when anchored to a solid surface so the technology has sought to increase the surface area available by, for example, providing helical inserts to fit into conventional culture bottles. One of the latest is to grow cells on 'microcarriers', tiny particles (about 200 mm diameter) on which cells attach and divide. The surface area afforded by 1 kg of microcarriers is about 2.5m^2 and the space taken up (a prime consideration in commercial practice) is very economical. This method combines the ease of handling cell suspensions with a solid matrix for the cell to grow on.

Identification of Viruses

Serology or the use of antibody

Without antibodies there would be no virology since there is no other simple way distinguishing between morphologically similar viruses such as the three different polioviruses. Antibodies are proteins produced by cells of the immune system of higher vertebrates in response to foreign materials (antigens) which those cells encounter. The antibodies are secreted into the body fluids and are usually obtained from the fluid part of the blood (antiserum) which remains after cloning has removed cells and clotting proteins.

Any method of detecting or measuring virus can be used such as inhibition of haemagglutination or immune electron microscopy where after reaction with specific antibody viruses can be seen in aggregates. Alternatively, antibody can be employed to detect viral antigens in the infected cell. When the cell is alive, antibodies are excluded and will therefore react only with antigens exposed on the surface of the cell. This permeability barrier is destroyed by 'fixing' the cell in acetone which enables antibody to localize antigens in the cytoplasm and nucleus. Antibodies are 'tagged' before use with a marker substance and hence can be detected *in situ*. Tags such as fluorescent dyes can be seen by UV microscopy, enzymes (peroxidase, phosphatase) which leave a coloured deposit on reaction with substrate can be seen by light microscopy, radioactive substances can be detected by deposition of silver grains from a photographic emulsion and electron dense molecules (e.g., ferritin, an iron-containing protein, or gold are visualized by electron microscopy.

Enzyme-lined antibody is now commonly used in a quantitative assay called ELISA (enzyme-linked immunosorbent assay). In the example the assay is being used to measure the amount of antibody present in an unknown sample but it can also be used to measure antigen. The couloured product is proportional to the amount of specific antibody bound and can be measured spectrophotometrically. Such assays can easily be automated to deal with large numbers of samples.

Western blotting

When antibodies were first used as a probe for macromolecules separated by gel electrophoresis it seemed logical to follow the earlier precedents of southern and northern blotting by calling the technique 'Western' blotting. For this procedure virus proteins are totally denatured and understand in terms of viruses which are made up of a number of identical repeating sub-units. For these viruses neutralization follows 'pseudo single-hit' kinetics where many antibody molecules combine with a virus particle and yet do not neutralize it, but the addition of one more antibody molecule performs the *coup de grace* and results in the loss of infectivity. In other words:

Virus + antibody → infectious (virus-antibody complex)

+ antibody → non-infectious (virus-antibody complex), i.e. neutralization

This explanation becomes clearer in the light of the concept of 'critical antigenic sites only these are relevant to neutralization. Thus combination of one molecule of antibody to a single antigenic site of phage T4 located on the base plate causes neutralization. Other sites are essential structural elements but are non-critical: thus, though antibody is raised against them and combines with them, neutralization does not occur. Other, and perhaps the majority of viruses have multiple critical sites and neutralization occurs only when sufficient of these are combined with antibody. The attachment of subneutralizing amounts of antibody to virus particles can be demonstrated by using radioactively labeled antibody. In the case of influenza virus, there are both critical and non-critical sites on the haemagglutinin molecules. The neuraminidase protein possesses only non-critical sites, as antibody to neuraminidase has no neutralizing activity.

In all neutralization reactions a small proportion of the virus remains infectious as a 'persistent fraction'. The virus particles in the persistent fraction are not genetically different from the neutralized ones but are found inside aggregates of particles which protect them from contact with antibody.

Neutralization is not the only way in which viruses can be inactivated by antibody' other factors can be involved such as the complement system or cells which have receptors for the COOH-terminal (F_c) part of the attached antibody molecules. Under these circumstances even non-neutralizing antibody can inactive infectivity.

Identification by restriction endonuclease mapping

Neutralization tests are indeed simple but can be slow as their usefulness depends on the time taken by the virus to kill a detectable number of cells; this can take up to several weeks. Such a situation is far from ideal for diagnostic virology and a recent development which aims to overcome the problem is to compare the patterns of DNA fragments produced by restriction endonucleases. The patterns are determined by a nucleotide sequence and are unique to a particular serotype. What happens now is that a newly isolated virus is assigned to its family by the old technologies—by reference to its cytopathic effects, its morphology by electron microscopy, or by immunofluorescent staining with antibody against a group specific antigen. Then virus is purified, and using the new technology, its DNA is extracted and digested with one or more restriction enzymes. The resulting fragments of DNA are separated according to size and identified by comparison with know standards.

This methodology is currently in transition from the research laboratory to the routine clinical situation. It is rapid, no more expensive than the neutralization test and can be performed on the virus produced by a single culture tube. One important aspect is that the method permits the study of some of the many strains of virus which can still not be cultivated in any way in the laboratory.

Southern and northern blotting

Until recently one of the major problems in cell biology was the difficulty of identifying a particular nucleic acid sequence amongst a vast excess of their sequences. For example, although DNA can be isolated using appropriate deproteinization procedures from infected cells and analyzed by gel electrophoresis, any staining technique will visualize all nucleic acid molecules, and the viral DNA will be obscured by the far larger amount of cellular material present. The problem was overcome by Ed Southern in Edinburgh who denatured and separated nucleic acids *in situ*, blotted them onto a cellulose nitrate filter and hybridized them to labeled nucleic acid sequence specific for the virus. Such a probe reacts only with the nucleic acid to which it is complementary and no other. The technique bears Southern's name and a subsequent procedure which detects RNA in a similar manner with labelled DNA probes was whimsically called Northern blotting. In this way the presence of viral genomes can be detected even when it is not possible to cultivate the virus or when the virus is latent and not actively multiplying.

Molecular probes and non-hazardous labeling

Short nucleic acid sequences (30-200 base pairs), prepared by *de novo* synthesis or isolated by recombinant DNA technology, are used to locate viral genomes in cells or tissues. They can even show the location of virus in longitudinal sections of whole animals.

Again what was once the province of the research laboratory is rapidly adopted and adapted to diagnostic virology but the usual radioactive probes are unsuitable as they have short shelf-life and contribute to the radiation load of a routine laboratory. As an alternative, nucleic acid probes (and antibodies) can be biotinylated before reacting with the test substance. After washing away unbound probe the test material is reacted with streptavidin which binds with high affinity to biotin. Next biotinylated polymers of intestinal alkaline phosphatase are added which bind to free sites on streptavidin. Incubation then with 5-bromo-4-chloro-3 indolyl phosphate and nitro blue tetrazolium deposits a blue coloration at the site of the original reaction. Nucleic acid sequences in the range 1-10 pg can be detected.

3

MOLECULAR VIROLOGY

The fields of molecular biology and virology have been interdependent for more than four decades. Molecular methodologies have transformed the study of viruses, viral pathogenesis, and viral diagnostics. Molecular biology itself has been changed by discoveries in the field of virology. Early viral studies helped establish the central dogma of molecular biology: genetic information residing in DNA is used to produce messenger RNA, which in turn is used to make specific proteins. The discovery of reverse transcriptase in two separate viral systems also served to disrupt the central dogma. Today, the molecular biologist uses a toolbox filled with viral artifices, from ligases to reverse transcriptases, DNA modifying enzymes to powerful transcriptional enhancers. It is not surprising therefore, that molecular methodologies are being used to diagnose viral illnesses.

Quality assurance programs constitute an integral part of any diagnostic laboratory and can account for 10-25% of the laboratory's consumable supply costs. Because of the exquisite sensitivity of nucleic acid amplification procedures, quality assurance programs in the molecular virology laboratory are more complicated (and expensive) than those used in other laboratories. The important elements of the molecular virology laboratory's quality assurance program are included in this chapter. However, quality assurance programs are not static, and they will continue to evolve as newer technologies move into the diagnostic laboratory.

SPECIMEN QUALITY

Specimen quality is the foundation on which all other procedures are built. Collecting the correct specimen by an appropriate method at the proper time during infection significantly enhances the probability that the laboratory will detect the virus, and improves the clinical relevance of the laboratory result. Poor quality specimens can undermine the value of even the best molecular diagnostic procedures. Because specimen collection is so critical, laboratories must establish specimen handling and adequacy standards that must be met before molecular diagnostic procedures are performed. Laboratories must also assume responsibility for training clinicians, nursing personnel, and laboratory technicians in the proper collection and handling of these specimens.

Specimen storage time and temperatures also have a significant impact on nucleic acid recovery and the efficiency of subsequent diagnostic procedures. Red blood cell lysis can influence polymerase chain reactions (PCR) by inhibition of *Taq* DNA polymerase by heme. Heme can also damage DNA at the elevated temperatures used in PCR. In addition, granulocyte lysis can release proteases and nucleases that degrade viral particles and nucleic acids. Cuypers *et al.* (1992) reported a 1000- to 10,000-fold reduction in hepatitis C virus (HCV) RNA concentrations when whole blood and serum were stored at room temperature for 14 days. However, nucleic acid degradation occurred even more quickly when whole blood was stored at 4°C. This difference was probably due to increased granulocyte lysis at 4°C than at room temperature.

Minimizing exposure to blood samples to temperatures ≥23°C is also important because specimens stored at these temperatures have decreased DNA yields. For these reasons, the laboratory must establish storage and transport standards for all specimens. The laboratory should also have a protocol for handling (or rejecting) specimens that do not meet these standards.

Once specimens are received in the laboratory, they should be held in their original containers. Pouring off or splitting specimens provides an opportunity for the introduction of extraneous nucleic acids that could produce a false positive result in a clinically negative patient.

Spectrophotometric evaluation and quantification of nucleic acids are often done to evaluate specimen adequacy in the molecular diagnostic laboratory. The most common spectrophotometric procedure involves determining the absorbance of the extracted specimen at 260 and 280 nm. The A_{260}/A_{280} ratio of highly purified DNA will approach 2.0; in most laboratories, samples with ratios below 1.7 are considered inadequate for further testing. The absorbance of the extracted specimen should also be determined at 230 nm because contaminants such as salts, sodium dodecyl sulfate (SDS), and some amino acids have peak absorbances at or near this wavelength. A_{230}/A_{260} ratios should be 0.3-0.5. Ratios that are substantially different from these values may indicate contamination with unacceptable levels of proteins and/or salts. Although useful and easy to perform, optical density measurements have some limitations. For instance, phenol strongly absorbs light at 270 nm, so the presence of phenol in the nucleic acid preparation can distort the A_{260}/A_{280} ratio. In addition, free ribonucleotides and deoxyribonucleotides absorb light at 260 nm; their presence can also falsely elevate the A_{260}/A_{280} ratio.

Careful analysis of nucleic acid yields and absorbance values can help the laboratory establish acceptable ranges for each specimen type. Specimens with values outside these ranges may not provide accurate diagnostic information and should be "*cleaned*" with 70% ethanol, re-extracted, or rejected. However, there may be compelling reasons for testing specimens that would normally be rejected. In these cases, an "*inadequate specimen*" comment should be formulated and included in the laboratory report.

Facilities and Equipment

The facility requirements of a molecular virology laboratory depend on the type and number of procedures performed in the facility. Laboratories performing nucleic acid amplification procedures must have at least two rooms or dedicated areas, whereas laboratories that do not perform these procedures can function in a single room. The molecular diagnostic laboratory must accommodate four separate functions: reagent preparation, sample extraction, reaction set-up, and postamplification product detection. The ideal molecular diagnostic laboratory will have a separate room for each function. However, laboratories with limited space can separate these functions by establishing separate work stations in one or two rooms. Each work station should have dedicated supplies and equipment. Benchtop hoods with ultraviolet decontamination lights should be used for reagent preparation, sample extraction, and reaction set-up areas. More elaborate facility designs have been described for larger laboratories. Laboratory work surfaces must be smooth and constructed to withstand regular cleansing with dilute acid, bleach, and/or shortwave ultraviolet radiation.

Quality assurance programs must include periodic equipment checks designed to ensure that the equipment is functioning as expected. Laminar airflow hoods should have face airflow velocities >75 In ft/

min at the work opening. Laminar airflow hoods should also be checked by a certified inspector at least once each year. The intensity of the ultraviolet (UV) radiation in laminal flow and static air hoods should be checked at the workbench level at least every 6 months and the ultraviolet bulbs should be replaced when the UV intensity drops below acceptable levels. In rooms with automatic UV ceiling lights, the UV intensity should be checked at several locations in the room to ensure even coverage.

The electrical supply should be checked before installing a thermocycler for two reasons: (1) these devices generally have high amperage consumptions and (2) the cycling of these devices can cause line voltage fluctuations that could damage other equipment. Thermocycling devices should be maintained carefully and checked for performance as part of the overall quality assurance program. Block temperatures should be checked periodically to ensure that the temperatures attained in each block position correspond to the digital readout of the instrument display panel. Thermocyclers should be decontaminated periodically by UV illumination or by swabbing the blocks with a cotton-tipped applicator soaked in 10% household bleach, followed by water-soaked applicators to prevent corrosion.

Centrifuges should be equipped with aerosol contaminant rotors to protect personnel from aerosolized pathogens and to prevent cross-contamination of specimens. Alternatively, centrifuge tubes with compressible O-ring seals may be used. Speed checks should be performed and recorded every 6 months. In addition, the condenser coils on refrigerated centrifuges should be cleaned at least twice each year.

The voltage of electrophoresis power supplies should be checked every 6 months to ensure that the voltage output corresponds with the setting on the power supply. pH meters should be calibrated daily before the first use of the day, using pH standards on either side of the target pH. The pH of the standards should be recorded and these records should be reviewed periodically by the laboratory supervisor. Balances should be kept clean and protected from drafts. Balances should be calibrated every 6 months using National Institute of Standards and Technology (NIST) Class S weight standards. The results of all these calibration checks should be recorded and compared with established tolerance limits. Equipment with performance characteristics outside the tolerance limits should be removed from service until it can be recalibrated.

PIPETTES AND PIPETTING

The most frequent cause of false positive results is the introduction of foreign nucleic acids via contaminated pipettes. Air displacement pipettes often produce aerosols during pipetting operations that can contaminate the barrel of pipetting devices. The use of positive displacement pipettes or aerosol-resistant pipette tips can significantly reduce nucleic acid contamination. Positive displacement pipettes use a disposable plunger and a molded tip assembly that physically isolate the pipette barrel from the solution. Aerosol-resistant tips contain a hydrophobic microporous filter that is bonded onto the walls of the pipette tip. The microporous filter traps aerosols before they can contaminate the barrel of the pipette. These filters can also prevent contamination of the specimen when a contaminated pipette is used inadvertently. Aerosol-resistant pipette tips can be used with standard pipetting devices and may provide a more economical alternative to positive displacement pipettes. However, aerosol-resistant pipette tips that have microporous filters that are merely inserted (not bonded) into the tip may not provide an air-tight seal and will therefore provide inadequate aerosol protection. The quality and performance of each lot of aerosol-resistant tips should be checked as described by Dragon *et al.* (1993) before being placed into service. Lots that fail this test should be returned to the vendor.

A number of precautions must be taken during sample collection, extraction, and nucleic acid amplification steps to minimize cross-contamination between samples. Gloves are a common source of cross-contamination and should be changed frequently. In addition, microcentrifuge tubes should be brought to room temperature and subjected to a quick "*pulse*" centrifugation before they are opened to minimize aerosolization and sample splatter. Microcentrifuge caps should be opened and closed carefully to minimize aerosolization of the contents. All nonsample components such as dNTPs, primers, buffers, and enzymes should be added to the tube before adding the sample. The "*hot-start*" PCR procedure described by Chou *et al.* (1992) has a higher potential for contamination because of the increased number of sample handling steps. However, even this procedure can be organized (e.g., by using wax plugs) to minimize contamination Finally, all nonessential tubes should be kept closed during sample addition to minimize cross-contamination.

Because accurate measurement of reagent volumes play a critical role in all molecular diagnostic procedures, each pipetting device should be serviced regularly and kept clean. All pipettes, including robotic pipettes, should be checked for accuracy and precision every 3-6 months or after being serviced.

Sample handling pipettes that are kept under ultraviolet lights to prevent nucleic acid contamination should be serviced and checked more frequently than other pipetting devices because ultraviolet light and ozone can degrade plastic parts and rubber seals. The accuracy and precision of adjustable pipettes should be determined at the high and low ends of the range and at the most commonly used volumes. Pipettes that fail the precision and accuracy checks must be recalibrated or replaced.

Biochemical Methods of Preventing Amplicon Carryover

Re-amplification of previously amplified sequences (amplicons) is the most frequent cause of false positive results in any nucleic acid amplification procedure. Several physical, chemical, and enzymatic nucleic acid inactivation methods have been developed to control amplicon carryover. Irradiation of PCR components with short-wavelength UV light was the first published method for inactivation of nucleic acids. Ultraviolet irradiation creates thymidine dimers and other covalent DNA modifications that destroy the ability of the nucleic acids to act as templates for further amplification. Although these methods are inexpensive and easy to perform, their inactivation efficiencies can vary widely.

Psoralens and isopsoralens have also been used to inactivate nucleic acids. Psoralens are planar tricyclic compounds that can intercalate between the base pairs of nucleic acids. When activated by longwave (300-400 nm) UV light, psoralens covalently cross-link pyrimidine bases and block template-dependent extension reactions. Isopsoralen compounds should be used whenever hybridization-based detection methods are employed because these compounds do not cross-link complementary strands and treated amplicons are capable of efficient and specific hybridization.

Hydroxylamine has also been used to inactivate nucleic acids after PCR amplification. Hydroxylamine preferentially reacts with cytosine residues to create covalent adducts that prevent base pairing with guanine residues. Hydroxylamine inactivation procedures are especially useful for eliminating short G + C products that are difficult to remove by other methods.

Enzymatic inactivation of amplified nucleic acids can be accomplished by incorporating uracil-*N*-glycosylase (UNG), a DNA repair enzyme, into the PCR reaction mixture. In this procedure, deoxythymidine triphosphate is replaced by deoxyuracil triphosphate (dUTR) in the PCR master mix. The resulting DNA amplicons contain deoxyuracil, the UNG substrate. If dUTP-containing contaminants are introduced into subsequent PCR mixes, UNG will remove the deoxyuracil residues from the DNA backbone. The resulting abasic sites are unstable and undergo alkaline hydrolysis at the pH and temperature conditions found in most PCR protocols. Hydrolyzed amplicons cannot serve as templates for further replication.

Although the previously mentioned approaches are effective for inactivating amplified products, they are not suitable for cleaning up spills or for routine cleansing of surfaces and equipment. Several studies have shown that dilute (10%) household bleach solutions provide the most effective means for inactivating nucleic acids on surfaces. Bleach solutions cause oxidative damage to target, primer, and amplicon nucleic acids and can be used to inactivate RNA and DNA. Prince and Andrus (1992) demonstrated that bleach was significantly more effective in inactivating DNA than hydrochloric acid treatment. Thus, regular use of freshly made 10% bleach solution should be an essential component of the quality control program of every molecular diagnostic laboratory. Because bleach solutions can damage finishes and corrode metal surfaces, it is important to rinse treated surfaces with autoclaved water or 70% ethanol to remove bleach residues.

Ultraviolet lights can also be used to control surface contamination. However, the UV source should be located within 30-60 cm of the working surface to provide effective cross-linking of DNA targets. Fairfax *et al.* (1991) reported that the use of UV lights in a biological safety cabinet for 1-2 hr was only marginally effective in reducing contamination by amplified DNA. However, overnight irradiation was effective in eliminating a 200-bp amplicon.

All nucleic acid amplification procedures, except perhaps the cycling probe technology, are susceptible to false positive results due to amplicon contamination. However, the effectiveness of the nucleic acid inactivation methods described here has yet to be determined with the newer amplification reactions.

Although no inactivation method can substitute for good laboratory technique, clearly, amplicon inactivation is an important part of every nucleic acid amplification procedure. Without an inactivation method, the newer amplification protocols will never become widely accepted in the clinical laboratory.

Protective Clothing

Universal precautions dictate that gloves, laboratory coats, and face shields should be worn whenever the potential exists for spilling or splashing body fluids. In the molecular biology laboratory, laboratory coats should be dedicated for use in each testing area and should remain in those areas, except when being cleaned. Because amplification products can be introduced into an assay by a shirtsleeve dragged across a contaminated benchtop, laboratory coats should be worn at all times while in the laboratory and should be removed when leaving the laboratory. Disposable sleeve protectors can also be used to provide additional protection when working in a laminar airflow hood.

Gloves should be worn to prevent contamination of the laboratory worker and to minimize the chances for contaminating the specimen with foreign nucleic acids and/or nucleases. However, gloves are easily contaminated and are an important source of cross-contamination, especially when handling microcentrifuge tubes. One method of minimizing glove contamination is placing a sterile 2 × 2 gauze over the cap when opening and closing microcentrifuge tubes (a separate gauze should be used for each tube). Aerosol contamination can also be a problem when uncapping vacuum-containing blood tubes. Glove contamination can be minimized if vacuum-containing tubes are uncapped under a gauze pad or with a disposable stopper remover. These procedures also provide some protection if the glass container is cracked or broken. Gauze pads and disposable stopper removers should not be reused.

Reagents and Glassware

Reagent preparation is an important part of any molecular biology laboratory. The reagent requirements of each laboratory will vary depending on whether the laboratory uses diagnostic test kits, commercially prepared reagent sets, or makes reagents from individual components. Whenever possible, all solutions should be autoclaved to minimize bacterial contamination and reduce potential degradation of enzymes and nucleic acids. However, contaminating DNA cannot be removed by autoclaving. Master mixes of nucleic acid amplification reagents and buffers should be aliquoted into single-use volumes to prevent contamination and to improve run-to-run reproducibility. Master mixes can be prepared, aliquoted, and stored under oil at –15°C for up to 2 weeks or at –70°C for months with minimal loss of activity. Meticulous record keeping is required; the laboratory must be able to trace all reagents used in a diagnostic test to the original vendor and lot number. Expiration dates must be established for all reagents; expiration dating must not be exceeded except in carefully defined cases (e.g., rare or expensive reagents). If expiration dating is to be extended, the laboratory must establish performance criteria and testing frequencies to ensure that the reagents are functioning as expected.

Reagent water quality should be carefully controlled because water is the largest single component used in the molecular diagnostic laboratory. Deionized or glass-distilled water should be used for all reagents. Water quality should be monitored and recorded daily to ensure that it is suitable for diagnostic procedures (≥10 megaohm-cm resistivity). Because reagent water should be sterile, most deionizing systems include a 0.2-μm filter at or near the dispenser. Flexible-tubing extending from the dispenser can serve as a source of bacterial contamination; care should be taken to position the tubing so that it drains fully after each use. Water sterility should be checked weekly using bacteriological media.

All volumetric glassware should be NIST Class A or calibrated against a Class A standard. Broken, chipped, or cracked glassware should be discarded. Washed glassware should be rinsed thoroughly to remove any residual detergents. Glassware should be checked for residual detergent using pH paper to determine the pH of representative pieces of wet glassware. Because most detergents are alkaline, the presence of detergent residue will cause an increase in the normal pH. Some nonionic cleaning agents cannot be detected by this method, so an indicator such as bromosulfone—phthalein must be used.

General laboratory glassware is often a source of RNase contamination that can plague laboratories testing for, or generating, RNA. In these laboratories, glassware should be baked at 250°C for at least 4 hr. Alternatively, glassware can be treated with a 0.1% diethylpyrocarbonate (DEPC) solution for 12 hr at 37°C. Residual DEPC must be removed from the glassware by heating to 100°C for 15 min or by autoclaving. Most molecular diagnostic laboratories use sterile disposable plasticware whenever possible to prevent contamination.

Procedural Controls

Every batch of reactions should contain positive and negative controls. Controls in each PCR procedure should be chosen carefully. Positive controls should amplify weakly to monitor the sensitivity of the reaction and to reduce the potential for amplicon contamination. Two types of negative (no target) controls should be used with each batch of samples: environmental controls and contamination controls. The environmental control is the first tube to receive an aliquot of the master mix and the last tube to be closed before temperature cycling. At least one contamination control (no target control) should be included for every 18-24 samples. A positive result in any of the contamination controls dictates that all reagents must be checked for amplicon or target DNA contamination. A positive result in the environmental control invalidates the test run and the specimens should be tested again.

In addition to positive and negative controls, an amplification control reaction can be set up for each specimen. Amplification controls are used to demonstrate specimen integrity and to protect against false negative results due to inhibitors of the amplification reaction. A resident target gene such as β-globin or the human leukocyte antigen DQα (HLA-DQα) can be used as a control for specimens that contain human genomic DNA. Alternatively, a separate specimen aliquot can be spiked with a low copy number positive control to demonstrate the ability of the reaction to amplify the target nucleic acids. Multiplex PCR procedures that simultaneously detect target nucleic acids and the co-amplification control sequences (e.g., HLA-DQα) are ideal for this purpose; when available, they are extremely efficient and cost effective.

Proficiency Testing

Proficiency testing is a way of life in the clinical laboratory. Proficiency tests provide an excellent means of determining the testing variability among laboratories (external proficiency tests) and within laboratories (internal proficiency tests). Formal external proficiency testing surveys are not currently available for the molecular virology laboratory. However, some laboratories participate in informal proficiency testing programs by sharing specimens and comparing results.

Internal proficiency testing is an important part of the molecular virology laboratory. Some type of internal testing is required under the Clinical Laboratory Improvement Act of 1988 regulations. Internal proficiency testing can be started by using leftover virology (VR) survey materials from the College of American Pathologists. In the long run, the supervisor should generate blind specimens for testing by the laboratory staff. Additional training may be required for personnel who consistently fail the internal proficiency tests.

4

Nucleic Acid Blotting Techniques

The detection of electrophoretically separated nucleic acids immobilized on membranes by molecular hybridization was first described for DNA by Southern (1975) and for RNA by Alwine *et al.* (1977). Southern combined the use of restriction enzyme technology, the resolving power of gel electrophoresis, and the ability of nitrocellulose to bind nucleic acids in high salt to create blots of denatured DNA suitable for hybridization to radiolabeled probes. An analogous procedure for the detection of specific RNAs, commonly referred to as Nothern blotting, soon followed from the laboratory of Stark. The ability to identify specific nucleic acid sequences within the genome or as part of the pool of transcribed RNAs has proven to be an invaluable tool to all molecular biologists.

Southern, Northern, and dot/slot blot procedures are currently widely used for detection of clinically relevant viral nucleic acids. The blotting-based procedures are highly sensitive, reproducible, and relatively inexpensive. In addition, Southern blotting of products of the polymerase chain reaction (PCR) is an important adjunct procedure used to increase further the sensitivity of PCR and to confirm the identity of the amplified DNA.

Since their introductions, the Southern and Northern blotting procedures have evolved with some general alterations and numerous case-specific ones. In this chapter, we discuss techniques to prepare and process nucleic acids from cells and tissues for use in filter hybridization assays. We also discuss a variety of techniques to separate nucleic acid pools to hybridization as well as the preparation of hybridization probes and methods for detection and quantification of hybridization signals. In addition, specific examples of these technologies are shown for the detection of nucleic acids from human immunodeficiency virus type 1 (HIV-1).

Sample Preparation

Procedures for the isolation of nucleic acids from virus-infected clinical samples are numerous and particular to both virus type and sample source (i.e., tissue, feces, urine, blood, cell, culture, etc.).

DNA Preparation

Procedures for preparing total DNA from tissues and cells in culture are abundant but generally very similar. Important features in common include rapid isolation and cellular lysis to protect DNA from endogenous nucleases and the enzymatic degradation of cellular proteins. Described here are two procedures that have been used successfully to obtain high molecular weight DNA from tissues and cultured cells; the first, a more traditional procedure, requires 2 days to complete whereas the second can produce DNA suitable for restriction digestion or PCR analysis in 2-3 hr.

Method (I)

1. (a) Harvest adherent tissue culture cells by trypsinization or scraping followed by centrifugation. Cells growing in suspension may be centrifuged directly. Wash pelleted cells twice in phosphate buffered saline (PBS) to remove medium and serum proteins. (b) Mince tissues as finely as possible. These samples may be frozen in liquid nitrogen and crushed with a mortar and pestle or disrupted by dounce homogenization. DNA yields from the former procedure are greater, but the latter produces sufficient material for several restriction digests even from small biopsy samples.
2. Resuspend samples in 10 mM Tris-HCl, pH 8.0, 25 mM EDTA, 100 mM NaCl. Add sodium dodecylsulfate (SDS) (to lyse cells and denature proteins) and proteinase K (to digest the denatured proteins) to final concentrations of 1% and 0.1 mg/ml, respectively. Incubate samples 12-18 hr at 37°C.
3. Extract digested samples twice with phenol:chloroform:isoamyl alcohol (25:24:1) and once with chloroform:isoamyl alcohol (24:1); precipitate with 2 volumes 100% ethanol. Recover white strands of DNA (and contaminating RNA) by centrifugation. Wash pellet with 70% ethanol and dry.

4. Resuspend pellet in 10 mM Tris, pH 8.0, 1 mM EDTA, 1 μg/ml DNase-free ribonuclease (RNase).
5. Repeat organic extractions described in Step 3, followed by ethanol precipitation in the presence of 0.5 volume 7.5 *M* ammonium acetate.

Method (II)

This procedure, described by Grimberg *et al.* (1989), was designed for rapid DNA extraction from whole blood but works well with cultured cells and tissues. High molecular weight DNA suitable for restriction enzyme digestion as well as PCR amplification can be obtained with few manipulations in 2-3 hr, making this a highly attractive technique for the molecular diagnostic laboratory.

1. Release nuclei by adding a Triton X-100-containing solution (0.32 *M* sucrose, 10 mM Tris-HC1, pH 7.6, 5 mM $MgC1_2$, 1% Triton X-100) to whole blood or pelleted cells. Harvest by centrifugation at 900 *g* for 5 min.
2. Resuspend nuclei in 10 mM Tris-HC1, pH 8.0, 10 mM NaC1, 10 mM EDTA 1 mg/ml proteinase K. Incubate for 2 hr at 65°C. Jeanpierre (1987) has demonstrated that the activity of proteinase K is far greater when heat-denatured proteins are the substrates for digestion; at 65°C the enzyme self-digests.
3. Use the resultant DNA directly, or subject it to RNase treatment and organic extraction, as described in Method (I).

RNA Preparation

As does isolation of DNA, preparation of intact RNA requires rapid separation of RNA from endogenous nuclease activity. RNase is a highly stable protein capable of surviving autoclaving; therefore several precautions should be taken to avoid RNase contamination. Gloves should always be worn when preparing RNA or reagents to be used in the preparation of RNA since skin is a significant source of RNase. Aqueous solutions should be treated with diethylpyrocarbonate (DEPC) which inactivates RNase. Although procedures for removing ribonuclease from glass- and plasticware are relatively simple, disposable pipettes and test tubes are ribonuclease free and require no prior treatment.

Several procedures are commonly used to prepare RNA from tissues and cultured cells. Many rely on guanidinium salts, powerful protein denaturants, to inactivate ribonuclease activity and lyse the cells. The frequently cited method described by Chirgwin *et al.* (1979), and modifications of it, produce high yields of undegraded RNA from sources rich in ribonuclease. However, this procedure requires many manipulations as well as an overnight ultracentrifugation run, making it less useful to the molecular diagnostic laboratory. Instead, the simple and more rapid procedure of Puissant and Houdebine (1990) is recommended. In this simple one-step procedure, total RNA (high yield and undegraded) is isolated free of other macromolecules using guanidinium thiocyanate, phenol, and chloroform at acid pH.

1. Prepare stock solution of 4 *M guanidinium* thiocyanate, 25 *mM* sodium citrate, pH 7, 0.5% sarkosyl in advance. This denaturing solution is stable for up to 3 mo at room temperature. Prior to use, add 2-mercaptoethanol to a final concentration of 0.1 *M*.
2. Homogenize fresh or frozen tissues at 4°C in 10 ml denaturing solution. RNA can be prepared from monolayer culture by adding denaturing solution directly to the dish.
3. In either case, transfer 5 ml material to a disposable polypropylene tube. Add the following, with vortexing after each addition: 0.5 ml 2 mM sodium acetate, pH 4.0; 5 ml-water-saturated phenol; 1 ml chloroform.
4. Centrifuge at 10,000 *g* for 10 min. Remove the aqueous phase and precipitate RNA with an equal volume of 100% isopropanol at — 20°C. At acid pH, DNA and proteins remain in the organic phase and at the interface.
5. Recover the RNA by centrifugation at 3,000 *g* for 10 min.
6. Vigorously vortex the pellet following the addition of 2 ml 4 *M* LiC1 to dissolve contaminating polysaccharides. RNA, which is insoluble, is again recovered by centrifugation at 3,000 *g* for 10 min.
7. Dissolve the RNA pellet in 2 ml 10 mM Tris, pH 7.0, 1 mM EDTA, 0.5% SDS and extract with 2 ml chloroform.
8. Centrifuge at 3000 *g* for 10 min. Collect the upper (aqueous) phase and precipitate with an equal volume of 100% isopropanol in the presence of 0.2 *M* sodium acetate, pH 5.0.

This entire process can be completed in under 4 hr.

Polyadenylated Messenger RNA Preparation

Approximately 1-2% of total RNA is messenger RNA (mRNA); the remainder is composed of ribosomal, transfer, and other assorted small RNAs. For some applications, particularly when the RNA species of interest is present at very low abundance, it is desirable to analyze only mRNA. Removal of nonmessenger RNA species increases the amount of mRNA that can be analyzed by Northern blotting. Messenger RNAs contain polyadenylated [poly (A)] tails that are not present in other RNAs. Under appropriate conditions poly (A)-containing RNAs bind to an oligo-dT cellulose column whereas other RNAs pass through. As in all RNA procedures, RNase-free DEPC-treated solutions are essential, and the use of sterile, disposable plasticware is recommended. Oligo-dT cellulose columns may be purchased from several sources or prepared as described here.

1. Prepare a slurry of oligo-dT cellulose in 1 ml 0.1 *M* NaOH and pour into a small disposable column. For 2 mg total RNA, use 0.5 g oligo-dT cellulose.
2. Wash the column with 5-15 volumes of water.
3. Equilibrate the column with 10-20 volumes of high salt buffer (10 mM Tris-HC1, pH 7.4, 1 mM EDTA, 0.5 *M* NaC1).

Oligo-dT cellulose chromatography

1. Incubate 2 mg total RNA dissolved in 1 ml elution buffer (10 mM Tris-HC1, pH 7.4, 1 mM EDTA) for 5 min at 65°C.
2. Add NaC1 to a final concentration of 0.5 *M* and load the denatured RNA onto the column.
3. Collect the eluant and reload onto the column. Repeat this step two additional times to maximize the binding of mRNA to the column.
4. Wash the column twice with 2 ml high salt buffer and 3 times with 2 ml low salt buffer (10 mM Tris-HC1, pH 7.4, 1 mM EDTA, 0.1 *M* NaCl).
5. Elute the RNA with 2 ml elution buffer. To obtain highly pure poly(A) + RNA, the selection procedure (Steps 1-4) should be repeated a second time.
6. Recover the RNA by precipitation in 2.5 volumes of ethanol in the presence of 0.3 *M* sodium acetate.

Nucleic Acid Quantification

Quantification of DNA or RNA in solution is easily accomplished by ultraviolet (UV) spectrophotometry. Concentration and purity are determined by measuring the optical densities of the purified nucleic acid at 260 nm and 280 nm (A_{260} and A_{280}). Nucleic acids exhibit an absorbance maximum at 260 nm; proteins exhibit a maximum at 280 nm. The A_{260} of a 1 mg/ml pure DNA solution is 20; that of RNA is 25. The concentration can be determined by comparing the A_{260} of the nucleic acid sample with that of a 1 mg/ml solution. Highly pure nucleic acid solutions have an A_{260} to A_{280} ratio > 1.8; less pure preparations have a ratio < 1.8. The latter should be repurified before use.

Southern Blotting

Preparing DNA for hybridization to specific probes requires three steps: (1) restriction endonuclease digestion, (2) agarose gel electrophoresis, and (3) Southern transfer.

Restriction Digestion

Restriction endonucleases may be purchased from many sources and are often supplied with a concentrated reaction buffer (usually 10×). Activity of the restriction endonuclease is described in units where 1 unit is the amount of enzyme that completely digests 1μg of a defined DNA substrate in a given amount of time, usually 1 hr. DNA should be digested with several enzymes that produce fragments that are diagnostic for the virus in question. Digest 10 μg DNA in 1× restriction buffer with 2-5 units of restriction endonuclease/μg DNA. Keep the volume of the reactions to a minimum since they will eventually be loaded into the wells of an agarose gel. In addition, the added enzyme should not exceed 10% of the total volume of the reaction. Enzyme storage buffers frequently contain glycerol which, at higher concentrations, may be inhibitory to the reaction. To obtain a detectable signal from a DNA sequence that is present as a single copy in the genome, maximize the amount of target sequence loaded onto a gel. Generally, 10 μg total cellular DNA/restriction digest is sufficient when standard probes (≥500 bp) are labeled to high specific activity (at least 10^9 cpm/mg) and all other conditions are optimized. If the DNA under examination is less complex, for example, purified viral genomic DNA, or the target sequence is present at greater than one copy per genome, proportionately less DNA may be used.

In addition to the samples to be analyzed, other digestions should be performed. If the virus of interest can be cultured *in vitro,* DNA prepared from purified virus stocks of known titer should be digested as well. Alternatively, cloned viral DNA can be used. Inclusion of serial dilutions of such digested DNAs defines the sensitivity of the assay and allows quantification of viral DNA in the sample.

In addition to the digested samples loaded on the gel, one lane should contain molecular weight standards to allow determination of the sizes of restriction fragments in the assayed samples. These standards are readily available from commercial sources.

Electrophoresis of DNA Samples

Digested DNA fragments are separated on the basis of size by gel electrophoresis. The most commonly used format, discussed here, is the horizontal agarose gel, but other formats such as vertical acrylamide may also be used. Horizontal electrophoresis chambers, in a variety of sizes and styles, and power supplies are commercially available from a number of sources. Several factors influence the migration of linear DNA fragments through the gel, including:

1. Size of the DNA fragment. Migration through the gel is inversely proportional to the $\log_{10}$ of the size in base pairs of a linear DNA fragment. Larger fragments are retarded while smaller fragments migrate more rapidly.
2. Agarose concentration. This is a critical determinant of the effective range of separation of DNA fragments. For a 0.5% agarose gel, this range is about 1 kilobase (kb)-20 kb; for a 1% gel, the range is 0.5 kb-10 kb. DNA fragments outside the effective range will not be efficiently resolved.
3. Electrophoresis buffer. The two most commonly used buffers are TAE (0.04 *M* Tris base, 0.02 *M* glacial acetic acid, 1 mM EDTA) and TBE (0.089 *M* Tris base, 0.089 *M* boric acid, 1 mM EDTA). Although DNA fragments migrate about 10% faster in TAE buffered gels than in TBE gels, the difference in resolution is negligible. Note that the buffering capacity of TAE is low, so the buffer requires replacement or recirculation during extended electrophoresis runs. This step is not necessary when using TBE.
4. Applied voltage. At low electric field strengths (<5 V/cm), DNA fragment migration is proportional to the applied voltage. Electrophoresis at elevated field strengths (>5 V /cm) results in proportionally faster migration of higher molecular weight species and loss of resolution. Generally, for maximum resolution, electrophoresis is carried out at 1 V/cm.
5. The intercalating dye ethidium bromide causes a slight reduction in the mobility of linear DNA but no significant loss of resolution. Inclusion of this dye in the gel enables visualization and photography of the samples and standards after electrophoresis since ethidium bromide fluoresces on exposure to UV light.

Gels are cast on a horizontal plate sealed along the sides and ends with tape. To form the wells, a template or comb is mounted vertically near one end of the plate 1 mm above the glass plate. Agarose at the appropriate concentration is melted in electrophoresis

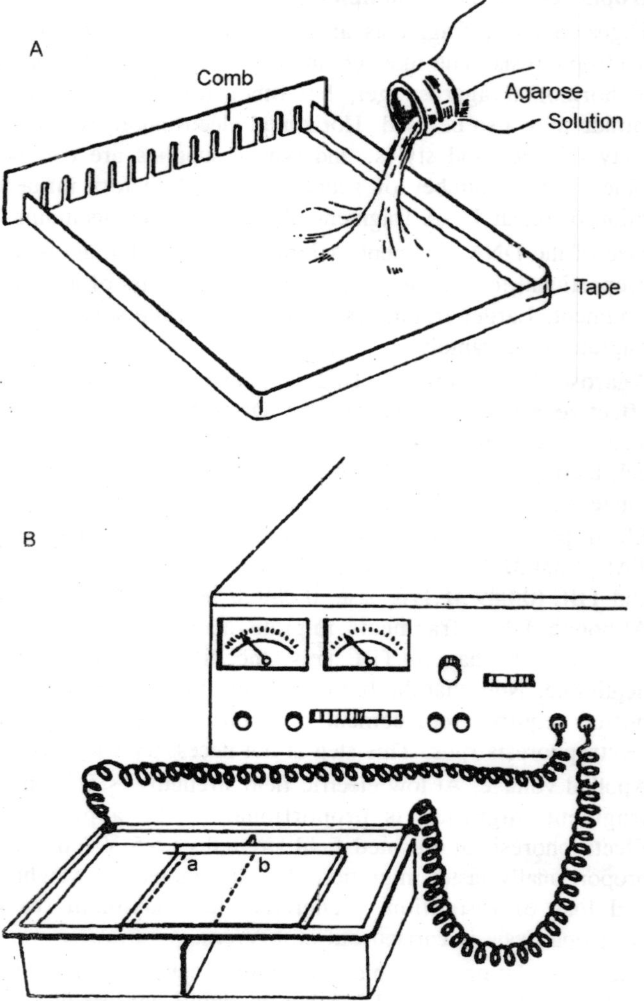

Fig. 4.1. Agarose gel electrophoresis. A–Melted agarose is poured into the tope-sealed gel plate with comb in place and allowed to solidify. B–The comb is removed and samples are loaded into the wells (a).

buffer (in a flask or bottle) and cooled to -60°C. Ethidium bromide is added to a final concentration of 0.5-1.0 μg/ml. The solution is mixed and poured onto the gel plates, avoiding air bubbles, and is allowed to solidify. After the comb and sealing tape are removed, the gel is placed in the electrophoresis unit; just enough electrophoresis buffer to submerge the gel is added. Samples and standards are loaded into the wells using a micropipetter. Several loading buffers are commonly used, which (1) ensure complete loading of the sample into the well by increasing the density of the sample and (2) provide dyes that migrate through gels at known rates. Sixfold concentrates of these buffers are often composed of 0.25% bromophenol blue (BPB) which comigrates with a DNA fragment of approximately 300 bp, 0.25% xylene cyanol FF which migrates as a 4-kb DNA fragment, and a reagent to increase sample density. This reagent may be any of the following dissolved in water, 40% sucrose, 15% Ficoll type 400, or 30% glycerol. These buffers are diluted to 1× in the sample which is now ready for electrophoresis.

The electrical leads are attached and voltage is applied so that DNA migrates toward the anode. The gel is run until the BPB and/or xylene cyanol FF have migrated to the appropriate position in the gel. The gel may be examined and photographed under ultraviolet light.

Transfer of DNA Fragments to Membranes

Transfer of DNA fragments to a membrane, as described by Southern, was achieved using nitrocellulose and upward capillary transfer. Since 1975, several modifications have been made that simplify, accelerate, and increase the sensitivity of blotting. The most significant improvement has been in the immobilizing membrane itself. Although nitrocellulose is still used in many laboratories, nylon membranes have several advantages including greater tensile strength, superior nucleic acid binding capacity, and retention of smaller fragments (as small as 50 by compared with 500 by for nitrocellulose). In addition, because of their greater strength, nylon membranes may be reprobed several times whereas nitrocellulose tends to fall apart with repeated probing.

Further, during transfer DNA adheres but is not covalently attached. To achieve attachment requires at most a brief exposure to a UV light source for some nylon membranes, compared with 2 hr of baking at 80°C under vacuum for nitrocellulose membranes. Clearly nylon membranes are the preferred material for immobilization. The only reported disadvantage of using nylon is an increased background with some types of nonradioactive hybridization probes. Two types of nylon membranes are available, neutral and positively charged. Either may be used in the following protocols, with only minor alterations for the neutral membranes. Described here are three procedures for transfer of DNA to nylon membranes.

Standard capillary transfer in alkaline buffer

1. Place the gel in a dish with 10 gel volumes of 0.25 *M* HC1 and gently rock. Continue for 10-15. min after the xylene cyanol FF and BPB become green and yellow, respectively. This treatment results in partial depurination and facilitates the transfer of DNA fragments larger than 5 kb. For fragments smaller than 5 kb, this step may be omitted.
2. Drain the dish and add 10 volumes transfer buffer; rock the gel until the dyes have returned to their original color. For a neutral membrane, the transfer buffer is 0.25 *M* NaOH, 1.5 *M* NaCl. Increase the NaOH concentration to 0.4 *M* for positively charged membranes.
3. The transfer stack is assembled properly. Place blotting paper on a platform that is larger and wider than the gel in a large dish so the surface of the platform is covered and the ends of the paper drape over into the dish.
4. Fill the reservoir with transfer buffer to a level below the top of the platform. Allow the wick to absorb the buffer and smooth out any air bubbles with a pipette.
5. Place the prepared gel on the platform and surround it with a waterproof barrier (Parafilm works well for this) so the wick around the gel is completely covered. This precaution prevents buffer flow from the wick directly to the paper towels above the gel. This short circuiting decreases transfer efficiency.
6. Wet a nylon membrane, cut slightly larger than the gel, in distilled H_2O for several minutes and place on top of the gel.
7. Place at least 12 pieces of blotting paper on top of the membrane and smooth out any air bubbles. Stack paper towels (4-5 cm) on top, place a glass plate on the paper towels, and set a weight on the glass. The weight should be sufficient to compress the towels, ensuring good contact between the gel and the membrane. The towels should not make direct contact with the reservoir or the wick.
8. Alkaline transfer is rapid and complete in 2-3 hr. Disassemble the apparatus and mark the membrane with pencil to indicate the position of the wells.
9. Rinse the membrane in 2× SSC (1 × SSC: 0.15 *M* NaC1, 0.015 *M* Na citrate, adjusted to pH 7.0) to neutralize the membrane and remove adhering bits of agarose.

If positively charged membranes are used, no further manipulations are necessary since DNA is covalently bound to the membrane under alkaline conditions. If neutral nylon membranes are employed, the DNA must be cross-linked to the membrane by exposure to UV light. The optimum time of exposure for cross-linking is dependent on the UV source and must be determined empirically.

Downward capillary transfer

This procedure reverses the direction of buffer flow in the transfer stack. In the standard procedure, the weight placed above the stack and the weight of wet paper towels can combine to crush the gel and slow the transfer process. The downward capillary transfer stack places little weight on the gel and results in reduction of transfer time to about 1 hr.

1. Prepare the gel and membrane as described for standard transfer.
2. Stack paper towels 3-4 cm high on a level surface.
3. Place 5 pieces of blotting paper cut larger than the membrane over the paper towels. Wet the uppermost piece with transfer buffer.
4. Place the membrane, then the gel on top of the blotting paper. Smooth out any bubbles with a pipette.
5. Surround the gel with parafilm, completely covering the paper towels.
6. Saturate 3 pieces of blotting paper, large enough to cover the gel, with transfer buffer and place on top of the stack.

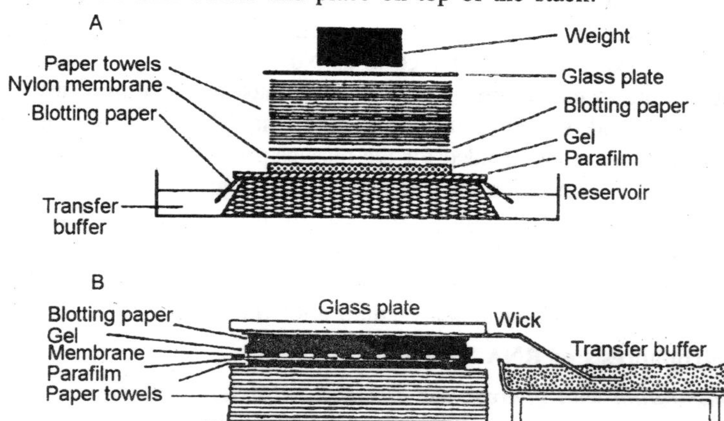

Fig. 4.2. Capillary transfer of nucleic acid from agarose gels to nylon membranes. A-Standard upward capillary transfer. B–Downward capillary transfer.

7. Saturate a large piece of blotting paper and place on the stack so the end drapes into a reservoir of transfer buffer.
8. Set a glass plate on top of the stack to prevent evaporation.
9. After 1 hr, disassemble the stack and prepare the membrane as described for standard transfer.

Pressure/vacuum transfer

An alternative to transfer of nucleic acids from agarose gels to nylon membranes by capillary action involves the use of either positive or negative pressure devices. The gel systems, electrophoresis, and postelectrophoretic gel preparation for pressure transfer are identical to the capillary transfer methods just described up to the point of transfer itself. At this point, the gel is placed on top of a piece of nylon membrane that is placed on a buffer- saturated piece of Whatman 3MM paper on top of a porous support. A sponge saturated with 10× SSC is then placed over the membrane/gel and the entire sandwich is sealed within the pressure chamber. Positive pressure is applied for 30-60 min to effect nucleic acid transfer. The nylon membrane is then marked for orientation, subjected to UV cross-linking, and processed for hybridization. The advantages of pressure blotting are primarily speed and uniform transfer efficiency across the gel. Positive-pressure blotting requires more effort to set up and involves the expenditure of several thousand dollars for equipment.

Northern Blotting

As can DNA fragments, RNA molecules can be separated on the basis of size by gel electrophoresis and can be immobilized on membranes by a process referred to as Northern blotting. To achieve tight binding to membranes, the RNA first must be denatured, usually by glyoxal and dimethyl sulfoxide or formaldehyde. Again, nylon membranes are recommended for the reasons previously discussed. The background hybridization may be significantly higher with some nylon membranes and may require the use of nitrocellulose. The two procedures described here work equally well with nitrocellulose or nylon membranes, but covalent attachment of RNA to nitrocellulose requires 2 hr of baking under vacuum after transfer.

Electrophoresis of RNA Samples

Denaturation with glyoxal and dimethyl sulfoxide (DMSO)

1. Prepare denaturation buffer in advnace [6 *M* deionized glyoxal, 75% DMSO, 0.1 *M* NaH_2PO_4, pH 7.0 (DEPC treated)]. Aliquot and store at –70°C. Deionized glyoxal may be prepared by passing

a 40% (6 *M)* solution through a mixed bed resin column until the pH exceeds 5.0.

2. Incubate 10-15 μg total cellular RNA in 15-20 μl denaturation buffer for 1 hr at 50°C. When rare messages are to be detected, up to 3 μg poly(A)+ RNA can be substituted. Samples may be lyophilized prior to addition of denaturation buffer if their volumes exceed 5 μl. Chill samples on ice and add loading buffer.

   As done when analyzing DNA samples, a dilution series of purified viral RNA from stocks of known titer should be included if possible. When using total cellular RNA, a duplicate sample may be included and 28S and 18S ribosomal RNAs may serve as size markers. Alternatively, or when using poly(A)-selected mRNA as samples, commercial preparations of RNA standards can be used. Subsequent to electrophoresis, lanes with standards can be trimmed from the gel, stained with 0.5 μg/ml ethidium bromide, and photographed under UV light. When using nylon membranes, the positions of the ribosomal RNAs can be visualized by exposure of the membrane to ultraviolet light. The ribosomal RNAs appear as dark bands on a bright background, and their positions can be marked on the blot with pencil.

   Prior to casting the gel, the gel plate, comb, and electrophoresis chamber should be washed with detergent and rinsed with H_2O and ethanol. The equipment may be further treated with 3% hydrogen peroxide to remove any contaminating RNase. The gel is cast and run in much the same way as a DNA gel with the following exceptions: (1) agarose, 1-1.5% depending on message size, is dissolved in 10 mM sodium phosphate, pH 7.0, which is also the running buffer, and (2) ethidium bromide, which reacts with glyoxal is omitted.

3. The gel is run at 3-4 V/cm until the BPB is 1/2 to 3/4 of the way through the gel. Recirculation of the electrophoresis buffer by peristaltic pump is essential to prevent the pH from exceeding to 8.0. Above this pH, glyoxal dissociates from RNA and artifactual migration of the RNA occurs. The gel is ready for transfer on completion of the run. No additional manipulations are necessary.

Formaldehyde denaturation

1. Add formamide and formaldehyde to final concentrations of 50% and 2.2 *M*, respectively, and running buffer diluted to 1× from a concentrated stock. Two running buffers are commonly used: (a) 10 mM NaH_2PO_4, pH 7.0, and (b) 0.02 *M* 3-(N-morpholino)

propanesulfonic acid (MOPS), pH 7.0, 8 mM sodium acetate, 5 mM EDTA. pH 8.0. Either may be used in this protocol. Denature RNA samples and standards for 15 min at 60°C.

2. Prepare the gel electrophoresis apparatus.
3. Melt agarose in water and allow it to cool to 60°C.
4. Add formaldehyde and concentrated running buffer to the agarose solution to final concentrations of 2.2 *M* and 1 x, respectively. Cast the gel in a chemical hood since formaldehyde vapors are toxic.
5. Perform gel loading and electrophoresis as previously described.
6. Remove the gel to a large dish, wash with several changes of distilled H_2O to remove formaldehyde, and equilibrate with 10x SSC. The gel is now ready for transfer of the RNA.

Transfer of RNA to Membranes

The transfer of RNA to nylon (or nitrocellulose) membranes can be achieved by following the procedures detailed for DNA transfer with the substitution of 10× SSC for the alkaline transfer buffer. High salt transfer is not as rapid as alkaline transfer and should be carried out for 12-24 hr to ensure complete transfer of RNA. RNA binds tightly but noncovalently to nylon and nitrocellulose under these conditions. Ultraviolet cross-linking is necessary for both charged and neutral membranes, as is baking at 80°C under vacuum for nitrocellulose blots.

Dot/Slot Blotting DNA and RNA

Dot and slot blotting are rapid, simple techniques for the quantification of RNA or DNA target sequences without prior electrophoretic separation. These methods differ only in the shape of the immobilized nucleic acid spot deposited on the membrane. As originally performed, nucleic acid was applied to a dry nitrocellulose filter and allowed to dry. The resulting "dots" were variable in size, making accurate estimates of target sequence concentration difficult. Manual spotting of samples has largely been replaced by the use of filtration manifolds.

The dot/blot apparatus, supports a membrane and is connected to a vacuum source. Fluid samples loaded into the slots are drawn through the membrane, which retains any nucleic acid in the sample. Filtration manifolds are manufactured by several companies in both dot and slot formats. Procedures for preparation of DNA and RNA dot/slot blots using positively charged nylon membranes are described next.

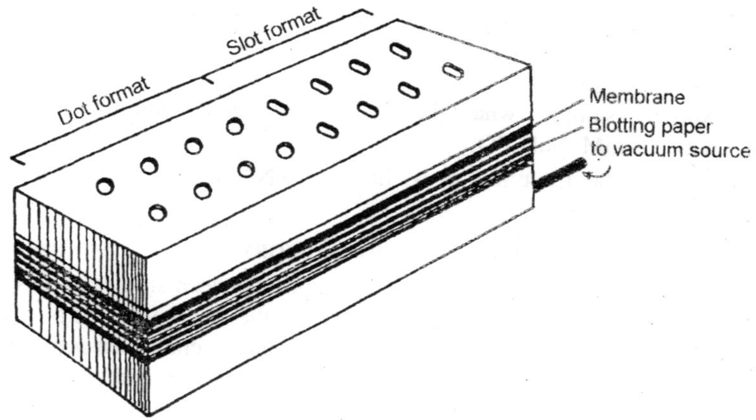

Fig. 4.3. Dot/slot filtration manifold. The apparatus is assembled with a membrane and blotting paper positioned as shown.

DNA Blots

1. In distilled water, wet one piece each of nylon membrane and blotting paper, precut to fit the manifold.
2. Assemble the manifold according to the manufacturer's specifications with the blotting paper supporting the membrane.
3. Prepare a dilution series of the DNAs to be probed as well as a control DNA in which the abundance of target sequence is known.
4. Denature the DNA by adding NaOH and EDTA, pH 8.0, to a final concentration of 0.4 *M* and 10 mM, respectively. Boil for 10 min. Immediately chill samples on ice.
5. Apply vacuum and add 100-500 μl water/well (depending on well size).
6. Load the samples when the water has been drawn through the membrane. Rinse each well with 200-500 μl 0.4 *N* NaOH. Remove the membrane from the apparatus and rinse in 2× SSC.

The alkaline treatment both denature the DNA and covalently attaches it to the membrane, thus eliminating the subsequent denaturation and neutralization required when neutral membranes are used.

RNA Blots

1. Assemble the apparatus as described for DNA blots.
2. Denature the RNA as described for sample preparation for Northern blots by using the glyoxal/DMSO or formamide/formaldehyde protocols, or by boiling for 5-10 min and then placing on ice.

3. Prepare a dilution series of RNAs to be probed, as well as controls.
4. Apply the vacuum and rinse wells with 200-500 μl 25 mM NaH_2PO_4, pH 7.0.
5. Add the samples when the wells have drained and wash twice with 25 mM NaH_2PO_4.
6. Remove the filter and cross-link the RNA to the membrane by UV irradiation.

Hybridization Theory

Molecular hybridization is the association between strands of DNA and RNA whose kinetics are governed by the degree of intra- and interstrand relatedness, solvent type, solvent ionic strength, temperature, and time. The basis for this aspect of molecular biology is the structure of DNA first proposed by Watson and Crick (1953). Their work showed that DNA was composed of two antiparallel strands of nucleic acid wound in a double helix with an external polyanionic phosphate backbone and an ordered internal association of the four constituent bases. These bases stack in the vertical plane of the helix by hydrophobic interactions and associate in the horizontal plane according to "*base pairing*" rules: purine bases, guanine (G) and adenine (A), must base pair with pyrimidine bases, cytosine (C) and thymidine (T), respectively. The rate-limiting step in molecular hybridization is the initial association between two antiparallel strands of nucleic acid that are capable of base pairing (complementary), termed "*nucleation*". Nucleation is enhanced when the concentration of the two complementary strands is high and when sufficient time is provided for the two strands to find one another. These two parameters, time and concentration of hybridizing strands, are critical to the kinetics of the reaction. In general, hybridization is favored by an increase in one or both of these parameters. In filter hybridization, detection of signal is enhanced by an increasing concentration of immobilized target nucleic acid. However, if quantification of varying amounts of immobilized nucleic acid is desired, the concentration of the probe must be in vast molar excess to the concentration of immobilized nucleic acid. Note that the kinetics of molecular hybridization are 7-10 times slower in filter hybridization than in solution hybridization, presumably as a result of steric effects.

The temperature at which two nucleic acid strands are 50% dissociated is termed the "*melting temperature*" (T_m). The T. for a given hybrid pair is a function of monovalent cation concentration, percentage of G:C base pairs, and length of the hybrid. Molecular hybrids are stabilized by the presence of monovalent cations since they mitigate the repulsive forces between the polyanionic backbones of two complementary nucleic acid strands. G:C base pairs, possessing three hydrogen bonds per base pair, increase the stability of a hybrid and, thus, the T_m to a greater extent than A:T base pairs, which contain only two hydrogen bonds per base pair. Hybrid length is not critical until the hybrid becomes quite small, as in hybridization using short oligodeoxyribonucleic acid probes. The effects of these factors on the T_m are described by the equation: $T_m = 16.6 \times \log [Na^+] + 0.4 (\% G+C) + 8.15°C - (600/\text{length in base pairs})$. For a 500-bp probe with a composition of 50% G+C in a typical hybridization buffer with a $[Na^+]$ of 0.825 *M*, $T_m = 16.6 \times \log [0.825] + 0.4 (50) + 81.5°C - 600/500 = 98.9°C$. The optimum hybridization temperature for filter hybridization is typically given by $T_H = T_m - 15$ to 30°C, in the 69-84°C range in the given example. Since these temperatures may be cumbersome to maintain, many investigators prepare hybridization buffers with a proportion of the nonaqueous solvent formamide. The effect of formamide on the melting temperature is given by $dT_m/d\%$ formamide = –0.7°C. Hence, for the hybridization just described, a buffer containing 50% vol/vol formamide would result in a T_m of 63.9°C and a T_H of 34-49°C with an average of 42°C. At these concentrations of formamide, RNA:RNA hybrids are more thermally stable than cognate RNA:DNA hybrids, which are more stable than cognate DNA:DNA hybrids. This effect will be of practical importance in discussions of RNA probes later in this chapter.

Base mismatch lowers the T_m of a long hybrid by approximately 1°C for every 1.5% of mismatch. These effects are magnified if the mismatches are clustered and in the middle rather than at the ends of the duplex. Given the relatively broad range for T_H, a small degree of probe—target mismatch can be tolerated in hybridization experiments. Larger degrees of mismatch or the use of mismatched oligodeoxyribonucleic acid probes are more problematic and demand an empirical adjustment of both hybridization and washing conditions. The rate of hybridization can be increased 12- to 13-fold by the addition of 10% dextran sulfate to hybridization reactions. However, dextran sulfate increases the background in filter hybridization and is both expensive and cumbersome to use. Hybridization is relatively insensitive to pH effects in the range of 5 to 9. Care must be taken, however, not to perform hybridization experiments with RNA probes or targets with a pH >8.0, especially at higher temperatures, to avoid base-mediated RNA hydrolysis.

Hybridization Probes

Probe Types

The choice of probes for molecular hybridization is guided by multiple concerns. Whereas the use of cloned plasmid inserts, restriction fragments, and complementary RNA (cRNA) probes is not dependent on the determination of the nucleotide sequence of the DNA of interest, this information is required for oligodeoxyribonucleic acid- and PCR-derived probes. The most sensitive molecular probes are uniformly labeled cRNAs since they combine a high ratio of labeled atoms/probe length, lack of a competing antiparallel probe strand, and the favorable hybridization kinetics of an RNA probe. The specificity (or stringency) of all hybridization probes is primarily a function of the hybridization and wash conditions. However, the use of whole cloned plasmids has a theoretically lower specificity because of the inclusion of plasmid sequences in the labeled pool that increase the probability of spurious cross-hybridization. This effect can be avoided by purifying the cloned insert away from the plasmid vector before labeling. A brief description of the commonly used techniques to prepare molecular hybridization probes, as well as a detailed protocol for each, is presented next.

Table 4.1. Characteristics of molecular probes

Probe type	*Sequence requirement*	*Sensitivity*	*Specificity*
Cloned plasmids	No	High	Medium
Restriction/PCR fragments	No	High	High
cRNA	No	Very high	High
Oligodeoxyribonucleotides	Yes	Medium	High

Probe Preparation

Random primed synthesis of linear DNA fragments

This technique, first described by Feinberg and Vogelstein (1983), involves the replacement synthesis of labeled deoxyribonucleic acid precursors into linear DNA fragments by a peptide fragment of the *Escherichia coli* DNA polymerase I (Klenow fragment). Circular plasmid DNAs must first be linearized with an appropriate restriction endonuclease prior to labeling. Restriction or PCR fragments, once purified, may be used directly in this method. It is especially critical that Mg^{2+} ions are removed from the DNA sample prior to labeling.

A random mix of hexanucleotide primers is utilized that allows the labeling of any double-stranded DNA template. As little as 25 ng template DNA may be labeled to produce probes $>1 \times 10^9$ cpm/μg with 70% incorporation of the labeled deoxyribonucleotide triphosphate precursor. These probes, once denatured, are then suitable for the detection of single copy genes and rare mRNAs in Southern and Northern blotting experiments, respectively, or for the screening of plaque and colony lifts.

1. Mix DNA of interest and 1-5 μg random hexamers into a total of 14 μl water or TE in a microcentrifuge tube.
2. Boil the DNA—primer mix for 2-5 min and plunge into ice.
3. Assemble the following in a separate microcentrifuge tube on ice:
   - 2.5 μl 0.5 mM d(G,T,C)TP, a mixture of 0.5 mM of each dNTP except dATP
   - 2.5 μl 10× Klenow/DNA polymerase I buffer (0.5 *M* Tris-HCl, pH 7.5, 0.1 *M* $MgCl_2$, 10 mM DTT, 0.5 mg/ml BSA)
   - 5.0 μl 3000 Ci/mmol [α-^{32}P]dATP (50 μCi)
   - 1.0 Al Klenow fragment (3-8 units)
4. Add 11 μl reaction mix from Step 3 to the denatured DNA—primer mixture.
5. Incubate at 37°C for 30 min or at room temperature for 2-16 hr.
6. Add 1 μl 0.5 *M* EDTA, 3 μl 10 mg/ml carrier tRNA, and 100 μl TE.
7. Extract with an equal volume of buffer-saturated phenol.
8. Pass the extracted aqueous layer through a Sephadex G-50 spin dialysis column.
9. Count 1 μl eluate in a scintillation counter.
10. If not used immediately, store the probe at -20°C.

Nick translation of recombinant circular plasmids

This technique, first reported by Rigby *et al.* (1977), is the oldest method of labeling hybridization probes that is still widely used today. Although initially described for the uniform labeling of circular plasmids, the nick translation reaction is suitable for linearized DNA fragments as well. The reaction involves the action of the enzymes DNase I and *E. coli* DNA polymerase I. DNase I is a 3' endonuclease that introduces single-strand scissions ("*nicks*") into the DNA duplex, resulting in free 3' hydroxyl ends. These 3' hydroxyl ends are then recognized by DNA polymerase I, which extends the nick through the combined action of its 5' to 3' polymerase and 5' to 3' exonuclease activities. Thus, the polymerase degrades the DNA strand proximally while synthesizing new DNA distally. This neck then progresses, or translates, along the duplex until the polymerase ceases to elongate the chain. Since DNA polymerase I is a poorly progressive enzyme, the length of DNA polymerized on any given run is approximately 400 nucleotides. It is critical to use DNase I sparingly and relatively high concentrations of dNTPs to prevent the resulting labeled DNA from being too small. Since *E. coli* DNA polymerase I is the only DNA polymerase known to possess a 5' to 3' exonuclease activity, it is the only enzyme suitable for the nick translation reaction. These concerns, coupled with the relative lability of DNA polymerase I preparations, have led to a gradual decline in the popularity of this labeling technique of DNA duplexes in favor of the random priming technique previously described. However, with careful attention to the reaction conditions, probe specific activities of 1×10^8 cpm/gg can be obtained.

1. Add the following to a 0.6-ml microcentrifuge tube on ice:
   - 2.5 μl 0.5 mM d(G,T,C)TP, a mixture of 0.5 mM of each dNTP except dATP
   - 2.5 μl 10× Klenow/DNA polymerase I buffer 0.5 *M* Tris-HCl, pH 7.5, 0.1 *M* $MgCl_2$, 10 mM DTT, 0.5 mg/ml BSA)
   - 10.0 μl 3000 Ci/mmol [α^{32}P]dATP (100 μCi)
   - 1.0 μl DNase I [freshly diluted 10,000-fold from a 1 mg/ml stock in 20 mM Tris-HCl, pH 7.5, 500 μg/ml BSA , 10 mM 2-mercaptoethanol]
   - 1.0 μl *E. coli* DNA polymerase I (5-15 units)
2. Add 0.25 μg DNA in a volume of 8 μl to the reaction mix to bring the final volume to 25 μl.
3. Incubate the reaction at 12-14°C for 15-45 min.

   Note: The precise incubation time should be determined empirically for a given batch of DNase I. Once the time to maximal incorporation of free label is determined, this time can be used in successive labeling reactions. Incubation beyond the point of maximal incorporation results in probe degradation with a resultant decrease in specific activity.
4. Stop the reaction by adding 1 μl 0.5 *M* EDTA 3 μl 10 mg/ml tRNA, and 100 μl TE.
5. Extract with an equal volume of buffer-saturated phenol.
6. Pass the extracted aqueous layer through a Sephadex G-50 spin dialysis column.

7. Count 1 μl eluate.
8. If not used immediately, store the probe at –20°C.

In vitro transcription

Many investigators have come to regard *in vitro* synthesized complementary RNAs (cRNAs) as the most useful molecular probes developed since their description by Melton *et al.* (1984) because of their combination of very high sensitivity and favorable hybridization kinetics. This technology exploits the use of plasmids containing the promoters for highly processive bacteriophage RNA polymerases (T3, T7, and SP6). These promotors are cloned just upstream from a sequence containing multiple restriction sites, facilitating the appropriate ligation of sequences of interest. The resultant recombinant plasmids are then linearized with an appropriate restriction enzyme that cleaves 3' to the cloned insert, before performing "run-off" transcription of the cloned sequences with bacteriophage RNA polymerase and labeled ribonucleotide triphosphate precursors, resulting in the production of cRNA. The plasmid DNA is then selectively degraded with RNase-free DNase; the cRNA probe is purified by phenol extraction and spin dialysis. These probes, when prepared using the protocol given here, yield specific activities on the order of 10^9 cpm/μg. These extremely sensitive hybridization probes should be used within 72 hr of synthesis since they undergo rapid radiolysis. More stable but less sensitive probes may be made by reducing the concentration of labeled radioactive precursor in the reaction.

1. Add the following in order to a 0.6-ml microcentrifuge tube at room temperature (spermidine—DNA complexes precipitate on ice):
   - 4.0 μl 5× transcription buffer (200 mM Tris-HCl, pH 7.5, 30 mM $MgCl_2$, 10 mM spermidine, 50 mM NaCl)
   - 2.0 μl 100 mM DTT
   - 0.8 μl placental RNase inhibitor (20 μl)
   - 4.0 μl 2.5 mM r(A,U,G)TP
   - 2.4 μl 0.1 mM rCTP
   - 1.0 μl form III DNA template in water or TE (0.2-1.0 mg/ml)
   - 5.0 μl 400 Ci/mmol [α^{32}P]rCTP (50 μCi)
   - 1.0 μl RNA polymerase (10-20 units)
2. Incubate at 40°C for 60 min.
3. Add RNase-free DNase (1 unit/μg DNA template).
4. Incubate at 37°C for 15 min.

5. Extract with an equal volume of buffer-saturated phenol.
6. Process through an RNase-free Sephadex (G-50 spin dialysis column.
7. If not used immediately, store the probe at –20°C.

5' End-labeling of oligodeoxyribonucleotides

The widespread availability of automated phosphoramidite technology to synthesize oligodeoxyribonucleotides has made the use of these reagents as molecular hybridization probes technically and financially feasible. Construction of appropriate oligodeoxyribonucleotide probes requires knowledge of the target nucleotide sequence. However, if this information is available, the creation of oligodeoxyribonucleotide probes avoids the subcloning step necessary for cRNA probes. If the target nucleic acid is not available in a laboratory for the production of nick-translated or random primed probes, an oligodeoxyribonucleotide hybridization probe can be prepared based solely on the knowledge of the nucleotide sequence of the target nucleic acid.

These reagents have been especially useful in the recovery of genes encoding proteins for which only amino acid sequence data was available. This information allows the construction of a family of oligodeoxyribonucleotides with a proportion of degenerate bases that correspond to all possible codons for the amino acid sequence. This set of "degenerate oligodeoxy-ribonucleotides" can then be used to screen bacteriophage cDNA libraries to purify the coding sequences for the protein of interest. These short molecular hybridization probes have also proven useful in the detection of small portions of target DNA or cDNA sequences amplified by PCR. The most common method of labeling oligodeoxy-ribonucleotides is the addition of a radioactive phosphoryl group from the gamma position of ATP to the free 5' hydroxyl group of the synthetic oligodeoxyribonucleic acid molecule, a reaction catalyzed by the bacteriophage T4 enzyme polynucleotide kinase. RNA, especially tRNA, is also a suitable substrate for this enzyme.

1. Add to a 0.6-ml microcentrifuge tube on ice:
   - 0.5 μl 20 *μM* oligodeoxyribonucleotide (10 pmol)
   - 2.0 μl 10× 5' end-labeling buffer (500 mM Tris-HCl, pH 7.6, 100 mM $MgCl_2$, 50 mM DTT, 1.0 mM EDTA, pH 8.0)
   - 5.0 μl 5000 Ci/mmol [γ^{32}P]ATP (50 μCi)
   - 11.5 μl deionized water
   - 1.0 μl T4 polynucleotide kinase (8 units)
2. Incubate at 37°C for 30-45 min.

3. Separate phosphorylated oligodeoxyribonucleotide from free label on Sephadex G-25 spin dialysis column. *Caution*: The use of a Sephadex G-50 column will result in a significant retention of labeled oligodeoxyribonucleotide probe on the spin dialysis column.
4. Count 1 μl eluate.
5. If not used immediately, store the probe at -20°C.

Note: Synthetic oligonucleotides are typically constructed with 5' hydroxyl groups. Therefore, they can be used as substrates for phosphorylation with polynucleotide kinase without prior dephosphorylation.

Filter Hybridization

Detection of target sequences in Southern, Northern, and dot/slot blots is carried out under essentially identical conditions. Three basic processes are involved: (1) prehybridization, which saturates nonspecific DNA binding sites on the membrane with random DNA and polymers; (2) hybridization, during which specific labeled probes are annealed to target sequences; and (3) washing, to remove unhybridized and imprecisely hybridized probe.

Numerous protocols have evolved since filter hybridizations were first described by Gillespie and Spiegelman (1965). The methods differ with regard to many factors including composition of prehybridization and hybridization buffers, hybridization temperature, and washing procedures.

Two protocols are presented here that are commonly used with nylon or nitrocellulose membranes.

Method (I)

1. Place the membrane into a heat-sealable bag and wet with a small amount of 2× SSC or prehybridization buffer (6× SSC, 5× Denhardt's reagent, 0.5% SDS, 100 μg/ml denatured sheared salmon sperm DNA; 50% vol/vol formamide may be included). A 50× concentrate of Denhardt's reagent consists of 1% each bovine serum albumin (BSA), polyvinylpyrrolidone, and Ficoll 400.
2. Decant the liquid, and add 15 ml prehybridization buffer for a 10× 15 cm blot. Proportionally less buffer should be used for smaller blots.
3. Seal the bag and submerge in a 65-68°C (42°C if formamide hybridization buffer is used) shaking water bath for 1 hr.
4. Single-stranded probes may be used without any pretreatment. Denature double-stranded probes before hybridization by boiling

for 5 min and immediately chilling on ice. Remove the bag from the water bath.

5. Decant the prehybridization buffer from the bag and add the same volume of hybridization buffer with probe. Hybridization buffer is identical to prehybridization buffer without Denhardt's reagent.
6. Reseal the bag, excluding any air bubbles, and return to the water bath for at least 6 hr or overnight.

Sensitivity and stringency of detection are determined by salt concentration and temperature of the buffer during the washing steps. Low stringency washes, with high salt at lower temperatures, increase sensitivity but may also increase background hybridization. High stringency washes, with low salt at higher temperatures, lower sensitivity but limit hybridization to highly specific probe—target interactions. The following protocol is for high stringency washing. Reduction in stringency may be made by appropriate alterations in temperature and salt concentrations.

1. Remove the membrane from the bag to a dish containing at least 200 ml 2× SSC, 1% SDS; agitate at room temperature for 15 min.
2. Decant the liquid and repeat the wash twice.
3. Wash twice for 15 min at 37°C in 0.1× SSC, 0.1% SDS, and twice for 30 min at 65°C in the same buffer.
4. Air dry the membrane.

Method (II)

This procedure, first described by Church and Gilbert (1984), is highly sensitive and yields blots with a high signal-to-noise ratio when using nylon membranes.

1. Place the membrane in a heat sealable bag.
2. Wet with water, decant, and add 15 ml hybridization buffer per 10× 15 cm membrane. Hybridization buffer is composed of 1% BSA, 0.5 *M* $Na_2HPO_4.7H_2O$, 1 mM EDTA, 7% SDS, adjusted to pH 7.0 with NaOH.
3. Agitate the membrane in the sealed bag for 5 min in a 65°C water bath.
4. Add probe, denatured as described if necessary.
5. Reseal the bag and incubate overnight at 65°C. In this procedure, stringency is determined solely by the temperature of the washes since salt concentrations remain constant. Stringency may be altered by increasing or decreasing the temperature appropriately.

6. Wash the blot twice for 10 min in at least 200 ml 0.5% BSA, 1 mM EDTA, 40 mM NaH_2PO_4, pH 7.0, 5% SDS at 65°C.
7. Wash 4 times for 10 min in 1 mM EDTA, 40 mM NaH_2PO_4, pH 7.0, 1% SDS at 65°C.
8. Air dry the blot.

Methods of Detection and Quantification

Signal Detection

The most common method for the detection of molecular hybridization signals is autoradiography. This technique involves the transfer of energy from the radioactive probe to silver grains contained on light-sensitive X-ray film. The bombardment of radioactive emissions onto the film causes these silver grains to darken that portion of the film on subsequent photographic development. Thus, the radioactive signal is transduced into a dark image on a clear background. The methods described in this chapter rely on the use of the strong β-particle emitting isotope ^{32}P. This isotope obviates the need for chemical signal enhancers in the sample to transduce emissions into higher energy photon emissions to expose the film. Thus, this particular type of autoradiography is termed "direct" autoradiography. Direct autoradiography is best performed at room temperature since this is optimal for film response. Exposure times are proportional to the radioactive signal intensity and are best determined empirically. The use of impregnating fluors, termed fluorography, although helpful for weaker β emitters such as ^{3}H, ^{14}C, and ^{35}S, offers no advantage for ^{32}P. However, the use of intensifying screens will boost film response times for ^{32}P probes roughly 10-fold. These image-enhancing screens work optimally at subambient temperatures (–70° to –80°C). It is recommended that most ^{32}P-probed Southern or Northern blot experiments be analyzed first by a short exposure to X-ray film in the presence of an intensifying screen at low temperature. Thereafter, optimal film exposures can be obtained with prolonged exposures with an intensifying screen at low temperature or without a screen at ambient temperature. The only real disadvantages of intensifying screens are slight loss of signal resolution due to emission scattering and the high cost of the screens themselves. The two most commonly used Kodak films are XAR-5 and XRP. The former is faster but of lower contrast than the latter.

Technologies have now been developed that allow more rapid detection of radioactive images. These systems either directly sample the radioactive emissions or scan the energy from the sample that has first been transferred to a storage phosphor screen. These signals are then used to create a digital image of the original sample. Both these technologies allow a 10-fold reaction in image production time compared with autoradiography at low temperature in the presence of an intensifying screen. However, the power of these imaging technologies is that they can perform quantitative signal analyses, as will be discussed later.

Autoradiography protocol

1. Perform all manipulations in a photographic dark room with safelight only. Handle X-ray film by the edges or with gloved hands since fingerprints will appear as images on the developed film.
2. Wrap the radioactive sample tightly with clear plastic film (e.g., Saran wrap).
3. Place the sample up against a piece of X-ray film contained in a light- tight cassette. If an intensifying screen is used, first place the screen in the cassette with the active surface facing up; then place a piece of film against the screen; then place the sample against the film before sealing the cassette. A second intensifying screen can be placed on top of this sandwich to further boost signal speed. However, the second screen's active surface must face the active surface of the first screen.
4. Expose the cassette at ambient temperature if no intensifying screen is used or at –70 to –80°C if an intensifying screen is used.
5. Remove the film from the cassette and develop using commercially available developer and fixative solutions according to the manufacturer's directions before rinsing well with water and air drying. Alternatively, an automated X-ray film developer may be used.

   Note: It is advisable to allow cassettes exposed at low temperatures to thaw at room temperature for 30 min before developing the film, to prevent film condensation which can lead to stray signals on the film.
6. If re-exposure is desired, the cassette, sample, and intensifying screens must all be warmed to ambient temperature and free from condensation to prevent the appearance of water marks on the subsequent autoradiogram.

Signal Quantification

Many investigators regard the use of Southern and Northern blot technologies as qualitative tools. They look for the presence or absence of signals or, at best, at faint, strong, and intense signal differences. Older methodologies to quantify autoradiographic images as well as the more recent imaging technologies previously discussed are available. The older methodology, quantitative densitometric scanning, involves the determination of the optical density of autoradiographic signals on X-ray film using a laser scanner. This technique requires rigorous attention to film exposure conditions and the simultaneous use of a series of known standards to determine the linear (dynamic) range of the assay. The X-ray film should be preflashed to an optical density of approximately 0.05 with a shielded light source before autoradiography to improve the dynamic range of the assay at low signal intensities. This technique, under the most ideal conditions and after multiple autoradiographic exposures and densitometer runs, is capable of a dynamic range of only 1.5-2 orders of magnitude. Despite these limitations, laser densitometers are affordable for many laboratories.

Innovations in image processing of radioactive samples involve the use of direct detection of radioactive emissions or the detection of radioactive energy stored in phosphor screens previously exposed to radioactive samples. These technologies are capable of quantifying radioactive samples with dynamic ranges of 3-5 orders of magnitude without the cumbersome constraints of laser densitometry. The price of these instruments puts this technology out of the reach of many investigators.

Examples of Blotting Technology

Southern Blot: Detection of HIV-1 RNAs of Negative Strand Polarity by PCR

The presence of negative strand HIV-1 transcripts produced by H9 cells acutely infected with the HXB3, RF, or MN strain of HIV-1. A reverse transcriptase—PCR (RT-PCR) approach was employed with primers specific for a 136-bp fragment of *gag* that can amplify cDNA of either positive or negative strand orientation, depending on the polarity of the primer chosen for the reverse transcription step. Detection and sequence specificity of the amplified DNA fragment was shown by Southern blotting and hybridization with an internal oligodeoxynucleotide probe. The dependence on RNA as initial template was shown by parallel reactions performed in the absence of reverse transcriptase. RT-PCR reactions performed in the absence of a specific primer during the reverse transcription step to control for the possibility of nonspecific priming by small DNAs contained in the RNA samples were uniformly negative (data not shown). Relative signal intensities in these experiments could not be interpreted since the assay was performed in a strictly qualitative fashion.

Methodology

1. Prepare RNA from acutely infected H9 cells by the RNAzol method.
2. Convert 6 ng RNA to cDNA and amplify using an RT-PCR technique.
3. Electrophorese 20% of the DNA contained in the RT-PCR reaction through a 15% agarose gel in TBE buffer at 6 V /cm for 1 hr. Electrophorese *Hae* III digest of πX174 in parallel to provide size markers.
4. Rock the gel for 20 min in denaturing solution (1.5 *M* NaC1/0.5 *N* NaOH); then for 20 min in neutralizing solution (1 *M* Tris-HCl, pH 7.4/1.5 *M* NaCl) at room temperature.
5. Set up a capillary transfer using a nylon membrane and 10× SSC; allow to transfer overnight.
6. Mark the nylon blot with indelible ink for orientation and subject to UV cross-linking.
7. Prehybridize in 6× SSC/0.1% SDS /1× Denhardt's/100 μg/ml herring sperm DNA for 30 min at 42°C.
8. Carry out hybridization in the same buffer supplemented with 1.5× 10^5 cpm/ml of ^{32}P-labeled oligodeoxyribonucleotide overnight at 42°C.
9. Wash the blot 3 times for 15 min with 1× SSC/0.1% SDS at 37°C, followed by a single wash for 15 min with 0.1× SSC/0.1% SDS at 37°C.
10. Remove excess fluid from the blot and wrap in Saran wrap prior to autoradiography.

Northern Blot

Analysis of the induction kinetics of HIV-1 RNA in U1 cells is shown. Whereas the steady-state levels of spliced RNAs, sized at 4.4 kb and 2.0 kb, were substantial in unstimulated Ul cells, the level of unspliced RNA (9.4 kb) was only twice that of the background radioactivity on the blot by radioanalytic imaging. The level of 2.0-kb RNA doubled after 2 hr of phorbol myristate acetate (PMA) stimulation but the amount of the 4.4-kb species did not change significantly from the basal level. As determined by quantitative image analysis, both the spliced transcript class levels increased after 4 hr of stimulation. Unspliced RNA also began to accumulate at this point. Further increases in all species occurred after 22 hr of stimulation. Unspliced RNA accumulated more rapidly than spliced RNA. No further accumulation of 2.0-kb or 4.4-kb RNA is seen after 28 hr of stimulation. Little additional accumulation of HIV-1 RNA occurred at time points between 28 and 53 hr of stimulation. However, cell viability began to fall below 80% after 48 hr. Quantification of the RNA induction kinetics from U1 cells indicated a 3-fold rise in total HIV-1 RNA over the time course. The 4.4-kb and 2.0-kb RNA size classes rose 3-fold, as well. Unspliced RNA accumulated to a level 25-fold higher than that of unspliced RNA in unstimulated U1 cells.

Methodology

1. Prepare RNA from acutely infected H9 cells by the RNAzol method.
2. Dilute RNA into loading buffer (50% formamide/7% formaldehyde/ 20 mM MOPS, 0.05% BPB), ensuring that the volume of loading buffer is equal to or exceeds the volume of RNA. Heat samples at 50°C for 15 min prior to being placed on ice.
3. Electrophorese samples at ≤6 V/cm until the BPB tracking dye migrates 80% of the distance to the bottom of the gel.
4. Photograph the gel.
5. Rock the gel for 20 min in denaturing solution followed by 20 min in neutralizing solution.
6. Subject the blot to capillary transfer blot for 8 hr.
7. Rinse the blot thoroughly in distilled water and UV cross-link.
8. Carry out prehybridization in 50% formamide/5× SSC/5× Denhardt's/20 mM Tris-HC1, pH 7.4/25 μg/ml yeast tRNA for 3 hr at 42°C.
9. Carry out hybridization in the same buffer supplemented with 1.5 $\times$ 10^5 cpm/ml of a ^{32}P-labeled cRNA probe overnight at 48°C.
10. Wash the blot twice for 15 mM at room temperature with 2× SSC/0.1% SDS, followed by 2 washes for 15 min at 50°C with 0.1× SSC/ 0.1% SDS.
11. Remove excess fluid from the blot and wrap in Saran wrap prior to autoradiography.

Slot Blot

Total cellular RNA, in descending amounts from various time points following phorbol ester induction, was immobilized onto a nylon membrane using a slot blot manifold. The membrane was then probed with $^{32}1^3$-labeled tRNA, washed, and subjected to autoradiography. All four dilutions revealed slightly different amounts of signal in the original RNA samples when scanned using a direct radioactive imager.

Methodology

1. Dilute total cellular RNA to 0.1 μg/ml using 15× SSC prior to serial 10-fold dilution.
2. Boil all dilutions for 5 min and then plunge into an ice-water bath for 2 min.
3. Load 50 μl of the cooled dilutions into the appropriate portion of a slot blot manifold containing a nylon membrane and draw through by gentle vacuum suction.
4. Rinse the wells 3 times with 100 μl 15× SSC.
5. Remove the nylon membrane from the manifold and cross-link with UV.
6. Carry out prehybridization in 1× SSC, 100 μg/ml salmon sperm DNA, 1× Denhardt's solution at 42°C for 2 hr.
7. Probe the blot with 2 × 10^6 cpm/ml of ^{32}P-labeled tRNA in hybridization buffer overnight at 42°C.
8. Wash the blot 4 times for 30 min in 1× SSC at 37°C, followed by a single wash in 0.1× SSC at 37°C, and subsequent autoradiography and digital scanning.

Southern Blotting and Polymerase Chain Reaction

Southern and Northern blotting with subsequent nucleic acid hybridization remains the primary method of detection for many viruses in clincial specimens. With the advent of PCR, a second very powerful tool for detection of low abundance nucleic acids became available to the molecular diagnostic laboratory. The high sensitivity of PCR makes it ideally suited to detection of viral nucleic acids in infected materials, where copy number per cell can be extremely low. PCR and Southern blotting are not mutually exclusive procedures. Southern blotting of PCR products is routinely done to increase the sensitivity of detection and to confirm the identity of the amplified DNA. The main barrier to PCR detection of viral genomes is the design of appropriate oligonucleotide primers based on knowledge of viral target nucleic acid sequences. In general, this obstacle is not insurmountable since the nucleotide sequences of many viral pathogens have been determined, and custom synthesized as well as premade primer pairs for virus detection are commercially available. When primers are unavailable, blotting and hybridization techniques can often satisfactorily identify virus-infected material.

5

Quantification of Viral Nucleic Acids

As antiviral therapy becomes more commonplace, the clinical laboratory will require routine methods with which to monitor the efficacy of therapy and the development of resistance to antiviral agents. Because chronic viral infections are often treated with prolonged courses of potentially toxic drugs, following the effect of the treatment at the level of virus infection in individual patients is often important. One of the challenges of the clinical virology laboratory is the quantification of virus in a sample. Most quantitative strategies involve molecular amplification of the target nucleic acid (target amplification) followed by detection of the amplified product with standard, and relatively insensitive, methods. Although target amplification techniques such as the polymerase chain reaction (PCR) are extremely sensitive, these methods require extraction and purification of the target materials and extreme care must be used to limit cross-contamination. Once the target has been amplified, it is difficult to relate the resultant number of copies to the physiological level of virus because the steps involved often have unknown or variable efficiencies.

Signal amplification methods do not alter the number of target molecules and are therefore more amenable to quantitative analysis. In addition, viral nucleic acids are measured directly in clinical samples; the resulting signal is thus directly proportional to the concentration of specific nucleic acid in the patient sample. Many techniques for labeling probes for Southern blots or *in situ* hybridization (e. g., fluorescence, radioactivity, chemiluminescence, electrochemiluminescence) can be considered signal amplification techniques, but these methods usually require high concentrations of target nucleic acids.

To increase the sensitivity of direct detection, branched oligodeoxyribonucleotides (bDNA) molecules were developed by Urdea and others at Chiron Corporation. These molecules are used as amplifiers to incorporate as many as 1755 alkaline phosphatase molecules onto each nucleic acid target molecule. The bDNA signal amplification system is highly reproducible, can utilize relatively crude specimens (no purification of target), and requires no special handling procedures beyond those needed for standard laboratory practices.

Description of the Procedure

Human immunodeficiency virus type 1 (HIV-1) RNA levels in plasma, hepatitis C virus (HCV) RNA in serum, and hepatitis B virus (HBV) DNA in serum can be quantified using a solid phase nucleic acid hybridization assay (Quantiplex) based on bDNA signal amplification technology. RNA or DNA is detected in a sample based on a unique solution-phase sandwich hybridization assay coupled with signal amplification employing bDNA. Multiple specific synthetic oligonucleotides (5-9, depending on the assay) are used for target detection and capture onto the surface of a microtiter well plate. Another set of target probes (18-39, depending on the assay) mediates the hybridization of the synthetic bDNA amplifier molecules to the target. The bDNA amplifier molecules have 15 arms, each of which can bind 3 alkaline phosphatase-labeled probes. After washing, alkaline phosphataselinked probes are hybridized, to the immobilized complex. Detection is achieved by incubating the complex with a chemiluminescent substrate and measuring the light emission. Since the number of target molecules is not altered, the resulting signal is directly proportional to the concentration of the target nucleic acid. The quantity of specific target in the sample is determined from a standard curve.

Components of the bDNA Assay

The bDNA assay differs from many other amplification procedures because the components are all supplied by the manufacturer (Chiron Corporation). Therefore, this chapter describes the assay and its components but not their synthesis, since synthesis of branched nucleic acids would be impractical in a virology laboratory. Specific instructions and troubleshooting guidelines are provided in package inserts and in training sessions. The bDNA assay consists of target probes that

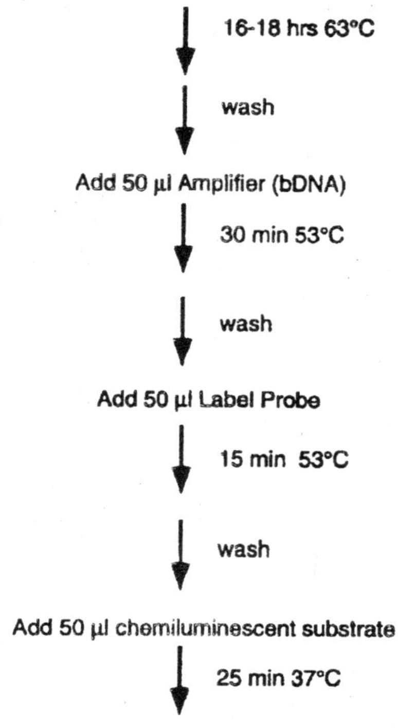

Fig. 5.1. Flow diagram of the laboratory procedure for the bDNA signal amplification assay for HCV RNA.

mediate capture, target probes that mediate attachment of amplifier, DNA amplifiers, alkaline phosphatase-labeled probes, and dioxetane, a chemiluminescent substrate.

Target Probes

The target probes are 50-base oligonucleotides, one portion (20-40 bases) of which is complementary to the target nucleic acid sequence while the second portion (20 bases) is used to capture the probe—target complex to an oligonucleotide-modified microtiter well plate or to attach the probe—target fragment to the bDNA amplifier. Several (5-50) target and capture probes can be used.

All the target probes are designed to hybridize to the most conserved regions of the target nucleic acids. For HCV, the probes are located in the 5' untranslated region and a portion of the core region; for HBV in conserved regions throughout the genome; and for HIV, in the reverse transcriptase *(pol)* region.

Branched DNA Amplifiers

Branched DNA amplifiers were initially synthesized by chemical cross-linking of oligonucleotides containing three alkylamine functions. However, the method was of limited utility because structural constraints precluded maximal probe binding. The use of "*branching monomers*" during the chemical synthesis of oligodeoxyribonucleotides proved to be a more flexible method. Branching monomers are phosphoramidite reagents containing at least two protected hydroxyl groups. In general, a primary linear fragment is synthesized and then tailed with several appropriately spaced branching monomers. Several simultaneous secondary syntheses are then initiated from the branch points. Branched DNAs containing several hundred nucleotides can be constructed in this manner. Large-branched oligonuceotides for signal amplification have been synthesized using a combination of solid-phase chemistry and enzymatic ligation methods. The amplifier–containing a maximum of 45 alkaline phosphatase probe-binding sites—is produced by synthesizing a bDNA with 15 branches that is then combined with a complementary linker that is in turn complementary to a branch extension or "arm," each of which has 3 binding sites for an alkaline phosphatase-labeled probe (3 sites × 15 branches = 45 labels). The amplifiers are assembled by ligation, then analyzed by capillary electrophoresis. On average, 13 of the 15 arms are incorporated into each branched oligonucleotide. Although the bDNA amplifiers are difficult to fabricate, they are easy to use.

Alkaline Phosphatase-Labeled Probes

Alkaline phosphatase probes are synthesized so part of the probe is complementary to a portion of the arms of the branched DNA. Three hybridization sites are located on each branch, yielding 45 alkaline phosphatase probes per bDNA molecule. Finally, alkaline phosphatase catalyzes the dephosphorylation of dioxetane, a reaction that produces visible light, the intensity of which is measured with a luminometer.

Molecular Standards

Extensive work has been done to determine accurately the number of target molecules in each of the standards used to generate a standard curve. Cloned and extensively purified target nucleic acids are quantified by several methods including phosphate determination, incorporation of [α-^{32}P]guanosine, and determining the absorbance of both intact and digested transcripts. Once the standards have been quantified, they are tested in the bDNA assay and the number of relative light units per attomole of target nucleic acid is determined. An "equivalent" is the amount of virus that generates the same amount of light as a single molecule of the standard. Any comparisons of absolute molecular sensitivity with other assays such as PCR would have to use exactly the same standard to compare numberical results.

Protocol for Hepatitis C RNA

Specific details are in the package insert.

1. Pipette 50 μl patient serum into each of two wells of a 96-well plate coated with capture probes. Pipette standards and controls into the appropriate wells. Add lysis diluent containing proteinase K, detergent and two sets of oligonucleotide probes. (Target probes with extensions mediate the capture of the target RNA to the microtiter well and target probes with extenders mediate binding of the target to the bDNA amplifiers). Incubate at 63°C overnight.
2. Wash plate with buffer A. Add bDNA amplifiers and incubate at 53°C for 30 min.
3. Wash plate with buffer A. Add alkaline phosphatase probes and incubate at 53°C for 15 min.
4. Wash plate with buffer A followed by buffer B. (The second buffer does not contain SDS, which will interfere with the enzyme-substrate reaction.)
5. Add dioxetane substrate. Incubate at 37°C for 25 min in the luminometer. Read relative luminescence.

Interpretation of Results

Clinical cutoff values for each of the assays were set by assessing the signal generated by a large number of negative samples and determining the level above which the assay could reliably distinguish a negative from a positive sample. Currently, the standard curve for HCV is 3.5×10^5 to 5.7×10^7 RNA equivalents/ml (the HCV assay uses 50 μl per well); for HBV, the curve is 7×10^5 to 4.5×10^9 DNA equivalents/ml (the HBV assay uses 10 μl per well); for HIV, the curve is 1×10^4 to 1.6×10^6 RNA equivalents/ml (the HIV assay uses 1 ml per well in a centrifuged pellet). The sensitivities are determined by studying populations of individuals known to be infected with each particular virus, as discussed subsequently. Sensitivity comparisons of populations are much more useful than molecular comparisons, since every laboratory uses different standards. The specificity of all the assays is greater than 98%.

Uses of bDNA Assays

Hepatitis C Virus RNA

HCV is the major cause of parenterally transmitted non-A, non-B hepatitis. Until the virus was characterized, diagnosis was made by exclusion of all other known causes of hepatitis. Antibody against HCV is found in over 80% of patients with well-documented non-A, non-B hepatitis. Chronic HCV is characterized by fluctuating alanine amino-transferase (ALT) levels and recognizable changes in liver histology, which may lead to cirrhosis and hepatocellular carcinoma. Therapy with interferon-α has been approved by the Food and Drug Administration (FDA) and can be effective in eliminating the virus. However, less than half of HCV-infected individuals respond to treatment, and relapse is common.

A bDNA-amplified test was developed for monitoring the level of the genomic RNA at the conserved 5' untranslated region and a portion of the core gene in individuals infected with HCV. In a study of chronic HCV patients, the bDNA assay detected and quantified the virus in 650/706 (92%) of those detected by RT-PCR. The quantity of HCV RNA in serum may be useful in monitoring both the disease state and the effectiveness of interferon therapy. Preliminary studies using bDNA to quantify HCV RNA during therapy indicate that there is a correlation between the level of RNA before the initiation of therapy and the prognosis. In other words, the greater the amount of RNA before therapy, the poorer the response.

Hepatitis B Virus DNA

Two studies were performed to determine the correlation of the HBV DNA assay with HBV e antigen (HBeAg) in infected patients. In the first study, with samples from patients in the United States, 53 of 56 specimens (95%) were positive. The second study, conducted with samples from Japanese patients, showed a 100% correlation (42 of 42) with HBeAg. In a comparison with the Abbott liquid hybridization assay, the bDNA assay detected 130/142 (92%),whereas the Abbott test detected 117/142 (82%). In HBeAg-negative, anti-HBe positive, HBsAg-positive patients, the bDNA assay detected 26/32, whereas a dot blot assay detected only 12. The quantity of HBV DNA has been shown to be useful in monitoring interferon treatment and in predicting response to therapy.

Human Immunodeficiency Virus RNA

In the HIV RNA assay, plasma samples (1 ml) are concentrated by centrifugation in a table-top refrigerated microfuge at 23,500 *g*, and then processed. In one study, 110 HIV antibody-positive samples were tested, and 79% were reactive; 93% of the 56 samples from patients with CD4 counts <200 were positive for HIV RNA, 67% of the 48 patients with CD4 counts of 200-500 were reactive, and 50% of the patients with CD4 counts >500 were reactive. The specificity of the test is 98.9%.

A significant number of HIV-infected individuals have high CD4 levels and high viral load. In a study using the bDNA assay to measure HIV RNA in the plasma of individuals enrolled in a large cohort study, researchers showed that patients with high levels of HIV RNA progressed to AIDS more rapidly than those who maintained low levels of RNA throughout the 30- to 60-mo follow-up period. Therefore, it may be useful to stratify patients by viral load rather than CD4 level for entry into clinical trials. The bDNA assay is currently being used in many laboratories to monitor changes in levels of HIV RNA for several antiviral treatment trials.

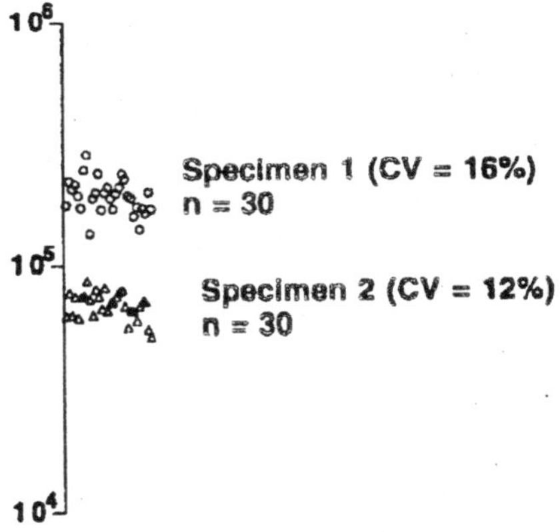

Fig. 5.2. Reproducibility of 2 samples run in 30 separate runs by branched DNA signal amplification for HIV RNA.

Sample Collection and Stability

Most of the assays have been developed for serum or plasma, which are not extracted before placement in the well. Extraction protocols using guanidine hydrochloride, sodium sarcosyl, and ethanol have been developed to quantify viral RNA (HCV or HIV) in lymph node tissue, liver tissue, and peripheral blood mononucleocytes.

Table 5.1. Sample collection and handling for quantitative bDNA assays for RNA

HIV

1. 5 ml *plasma* is required—this assures the ability to repeat any problem specimens and to do other tests if necessary (absolute minimum is 2 ml).
2. Blood should be collected in sterile tubes with anticoagulant—EDTA is preferred, acid citrate dextrose (ACD) or heparin is acceptable.
3. Remove plasma aseptically from the centrifuged collection tube within 4 hours of collection.
4. Dispense 1 ml aliquots (measured precisely) into sterile screw-capped tubes (1.5 ml Sarstaedt tubes with outer threads and rubber O-ring).

HCV

1. 0.5-1.0 ml *serum—this* ensures the ability to repeat any problem specimens and to do other tests if necessary (absolute minimum is 0.2 ml).
2. Blood should be collected in sterile serum separator tube (SST) with no anticoagulants.
3. Remove serum from the cells within 1 hour of collection by centrifuging the SST. (If SST is not used, serum must be removed from the clot within 1 hour.)
4. Remove serum aseptically from the centrifuged SST within 4 hours of collection and dispense into sterile screw-capped tubes (1.5 ml Sarstaedt tubes with outer threads and rubber O-ring).

Storage

1. Freeze immediately at —70 C after dispensing into sterile storage tubes.
2. After thawing, keep on ice during use.
3. Avoid repeated thawing and refreezing.

Since these assays are designed to be quantitative, it is especially important to preserve the physiological level of virus in each sample. RNA is very susceptible to degradation by the high levels of RNase in blood. Blood should be collected in sterile tubes; the serum or plasma should be separated from the cells within 4 hr of collection and then stored at -70°C. The bDNA signal amplification system has been used to detect and quantify a variety of viruses in clinical samples. Because of the simple steps and format, many samples (42-126 patient results) can be processed by one individual in 24 hr. The assay is very reproducible (interlaboratory CVs <30%) relative to other nucleic acid quantification methods. The dynamic range of each assay includes at least 90% of the detectable population without dilution. Because the viral nucleic acids remain at their physiological levels, the direct probe method is inherently quantitative and is not subject to the amplicon carryover problems associated with target amplification methods.

6

LABORATORY PRODUCTION OF VIRUSES

Although viruses have the ability to direct there own replication, they are totally dependent upon living cells for the machinery and materials to carry out the process. Thus, the routine laboratory cultivation virus requires adequate quantities of appropriate cells. Quite often particular viral agents will grow only in a specific type of cell from a certain host, such has monkey kidney or chicken embryo cells. The early attempts to grow animal virus were performed in susceptible animals. These included rhesus and other species of monkeys, hamsters, mice, and rates. Although these efforts were quite successful, laboratories were burdened by the need to purchase and maintain susceptible animals. Other means had to be found to meet the need of viruses and yet keep laboratory expenses within working limits. This chapter will describe some of the procedures currently used for cultivating viruses.

ANIMAL VIRUS CULTIVATION IN EMBRYONATED HEN EGGS

Embryonated eggs are eggs which have been fertilized by rooter. Have undergone a certain degree of embryonic development in the hen, and usually have been put into an incubator shortly after laying so that development can continue to the desire age, which might be 6 or 8 days. The introduction in 1931 of embryonated eggs as a suitable medium for the growth of viruses provided an invaluable means of obtaining large quantities of viruses for the preparation of vaccines and diagnostic regents, as well as viral studies. The first attempts to cultivate viruses by this these means was actually made in 1911, but it was not until the report of Woodruff, Goodpasture, and Budding in the United States and burnet in Australia that the chick embryo came into general usage.

The advantage if using such eggs include : (1) there availability virtually unlimited quantities, (2) the relative ease of growing and handling them, (3) the presence of a naturally constant environment within the confines of the egg's components—the embryo as if ordinarily comes from the hen is sterile, (4) the general inability of the embryo to produce antibodies against the agents used as inocula, and (5) the availability of eggs with relatively uniform genetic constitution from flocks which have been inbred for several generations. The many techniques used to chicken embryos in the cultivation of various viruses and rickettsia will not be discussed in great detail. However, brief descriptions of representative methods of inoculation and general precautions associated with these laboratory methods will be presented, primarily to emphasize the value of this simple experimental tool.

Generally, embryonated eggs for the cultivation of viruses are obtained from healthy flocks, free of both bacterial and viral agents. Commercially-reared chickens and routinely inspected for the symptoms of disease, and usually are inoculated against certain bacteria (e.g., *Salmonella pullorum*) and viruses (e.g., Newcastle disease virus—NDV). White shelled eggs, obtained from white Leghorn chickens, are more suitable than ones with dark or heavily-spotted shells primarily because the eggs' contents or more easily seen during *candling*, the examination of an egg with the aid of a light sources of sufficient intensity.

Embryonated eggs are collected shortly after laying and are refrigerated at temperature of approximately 10°C, for not longer then 10 days. Egg fertility decreases substantially after this time. Commercially available incubators generally regulated to maintain a temperature of about 37.5°C and equipped with devises to turn the eggs automatically several times daily are used to incubate eggs until the desired age of the embryo is reached. The relative humidity in such systems usually 55% to 65%.

When embryonic development has reached the desired point, the eggs are candled. Various types of candlers are available. A hand variety is most advantageous. This process is used to determine if the embryo is alive, and to locate the blood vessels and other parts of the animal. Living embryos usually exhibit spontaneous jerking movement (probably in response to the intensity of the strong light) and well-developed, translucent, floating blood vessels. Dead embryos are inmobile and their blood vessels are barely visible.

Before inoculation of the chick embryo, the surface of the eggshell to be used is sterilized with an iodine solution, which may are may not be followed by wiping with an alcohol solution. Sterile precautions are observed in the working area in which the inoculations are to be conducted. Drills, or other devices for boring into the shell, and surgical instruments generally are sterilized beforehand by standard methods such as boiling or autoclaving. In addition, the inocula used usually contain antibiotics to guard against bacterial and fungal contamination.

Several portions of the embryonated egg are utilized for viral cultivation, including the allantoic and amniotic cavities, the yolk sac the chorioallantoic membrane (CAM) and the embryo itself. The particular region utilized depends on the virus to be cultured, as certain viral agents are capable of proliferating well only in particular portions of the embryo. For instance, the introduction of the mumps and Newcastle disease viruses into the allantoic cavity results in the production of large quantities of these agents.

Table 6.1. Anatomical regions of the Chick embryo commonly used for virus cultivation.

Site of inoculation	*Age of the Chick embryos used*	*Representative viruses for which site is utilized*
Allantoic sac or cavity	12-13 days	Influenza, mumps, and Newcastle disease viruses
Amniotic sac or cavity (involves the amniotic fluid and cell of the amnion)	8-12 days	Eastern, Western, and Venezuelan equine encephalomyelitis viruses, mumps, and Newcastle disease virus
Chorioallantoic membrane (CAM)	9-14 days	Canine distemper virus, *Herpes-virus hominis* (the cause of fever blisters) variola (the causative agent of smallpox), vaccinia or cowpox
Yolk sac	6 days	Viruses of influenza, mumps, rabies, yellow fever, the Chlamydia causing psittacosis, trachoma, and lymphogranuloma venereum (a veneral disease) and rickettsia

The age of the chick is important, as particular regions must reach a certain stage of development to provide a suitable environment for virus proliferation. In order to introduce a virus-containing inoculum, a hole is bored in the shell. After the inoculation, this area is sealed with paraffin, transparent tape, or related material. The injected eggs are returned to the incubator until the particular reaction for which they were used occurs.

Signs of Virus Infection

Besides proving virus infection , the various embryo inoculation techniques are also used for : (i) studying the morphological features of viruses; (ii) isolating viruses from specimens; (iii) producing large numbers of vaccines; (iv) determining the effectiveness of drugs viruses (chemotherapy); and (v) investigating mechanism of infection.

Proof of virus infection in chicken embryos may be exhibited in several ways. Certain viral agents produce local lesions of varying sizes, shapes, and opacities, called *pocks*. However, it is not generally possible to completely identify a particular agent on the basis of this effect alone.

Additional signs of infection include: (1) the death of the embryo; (2) the demonstration of a hemagglutinating activities discussed further associated with the allantoic of amniotic fluids; and (3) the demonstration of the virus or related agents by certain techniques of light and electron microscopy. However, it is possible for there to be *no* obvious signs of viral infection.

Tissue Culture Cultivation of Viruses

The technique of tissue is almost as old as the specialization of virology itself. The development of modern animal cell culture is based on the contributions of many investigators, including the successful explanting of small tissue fragments in plasma clots by Harrison in 1907, the maintenance of vaccinia virus is explanted corneal tissue by Steinhardt, Israeli, and Lanbert in 1913, the demonstration by Parker and Nye in 1925 of viral replication (multiplication) in cultures of rabbit testis, and the development by Maitland and Maitland in 1928 of a technique causing cells to grow in fluid suspensions of small tissue fragments.

Despite the promising results obtained with tissue culture, many virologists were prevented from using the technique by contamination of preparations with bacteria and fungi, and the tedious, time-consuming procedures to prevent this hazard. Not until the introduction of antibiotics, the use of the enzyme trypsin to free cells from fragments of tissue so they can be grown in single-cells layers, and the formulation of excellent, defined-growth media for cells, did the methods of tissue culture gain routine acceptance by virologists for the cultivation of viruses and other agent requiring a living cell.

Although the tissue of cold-blooded animals and plants can be cultivated, the discussion which follows will be limited to the cultivation of mammalian tissues. According to most authorities, the final impetus which ushered in the era of tissue culture was provided by the *in vitro* cultivation of poliomyelitis virus in tissues other then nerve cells. This was achieved in 1945 by Enders, Weller, and Robbins, who adapted the suspended-fragment technique of the Maitlands and used human embryonic tissue for there investigations. Before then, polio virus was grown only in lower animals and nervous tissue. The destructive action of the virus originally referred to as "cytopathogenic" effect (CPE) but now designated "cytopathic," was clearly demonstrated. The three scientists received the Nobel Prize for their work.

Preparation of Representative Type of Animal Tissue Cultures

The preparation of animal tissue cultures generally involves both the various procedures by which cells are removed from animals and the procedures used to maintain them under suitable artificial conditions *in vitro*, so that they can serve as hosts for viruses or as sources of DNA, RNA, or other nucleic acids. The animal tissue must be provided with ample nutrients in their growth and maintenance media to keep their metabolic processes functioning in a desirable manner, and they must be kept free from contamination by unwanted microorganisms, such as mycoplasma and latent viruses. An unexpected hazard which confront individuals working with certain tissue cultures is the presence of viruses other than those are the subject of a particular investigation. The animals that are used as the source of tissue may be harboring a latent viral infection. When these tissues are grown *in vitro*, the viral agent may use its new environment to unleash some destructive effect, thus producing an extremely confusing picture for the virologist.

Glassware

The success of tissue cultures depends to a large extent upon the glassware (or plastic ware) used. Equipment of this type must be clear, nontoxic to cell, and chemically inter, not reacting with the ingredients of the media used to bathe the cells of the culture. Within in the past few years, a variety of disposable glass and plastic culture vessels has become available, including bottles, culture vessels such as tubes and flasks, Petri dishes and pipettes. Disposable items are generally clean and sterile and do not require preliminary washing. Nondisposable glassware generally must be pre-soaked in a mild detergent solution, washed repeatedly, rinsed, first in tap water and finally in glass-distilled water, quickly dried, and sterilized.

Media and solutions

The various media and solutions used in the culture of cells are commonly designated by the ingredients they contain, such as normal saline, phosphate-buffered saline, and chick embryo extract. Some solutions and media are simple, while others and quite complex. Examples of the former include balanced salt solutions and maintenance media. Synthetic or supplemented growth media are complex. Several types of balanced salt solutions are used in tissue cultures. Alone, these preparations do not support cell growth. They are employed: (1) to rinse the culture, thus freeing it from unwanted components, such as antibodies; (2) to dilute virus suspensions to be used for inoculation of cultures; and (3) as an important components of tissue culture maintenance or growth media. Solutions of this type generally consists of certain essential inorganic and organic compounds, a component such as sodium bicarbonate ($NaHCO_3$) to buffer solution (to stabilize the pH of the preparation), and an indicator such as phenol red to show changes in pH. All of these chemicals are dissolved in glass-distilled or deionized water. Many balanced salt solutions (BSS) are in common usage, including Earle's, Gey's and Hanks'.

Table 6.2. Typical components of common balanced salt solution.

Component	*Chemical Designation*
Calcium chloride	$CaCl_2$
Magnesium chloride	$MgCl_2.6H_2O$
Magnesium sulphate	$MgSO_4.7H_2O$
Potassium and sodium dihydrogen phosphates	KH_2PO_4; $NaH_2PO_4.H_2O$
Sodium bicarbonate	$NaHCO_3$
Sodium chloride	NaCl
Sodium monohydrogen phosphate	$NaHPO_4.7H_2O$
Glucose	$C_6H_{12}O_6$

Maintenance and Growth Media

The selection of medium is determined by the purpose for which a particular tissue culture is to be used. Cells that are to be transported or stored for a limited period of time are often kept alive in *maintenance media*. A medium of this type contains either small concentrations or no supplements such as serum or chick embryo extract, which are essential for cellular multiplication.

Growth media generally consist of balanced salt solutions with a variety of essential synthetic or natural additives. Examples of the latter include: (1) bovine and chick amniotic fluids : (2) sera from a variety of animals, such as calf, horse, human, ox, and rabbit; (3) embryo tissue extracts, (4) protein hydrolysates, such as tryptose phosphate broth and yeast extract. Note that the exact composition of these natural ingredient is not known. However, they are essential for the well-being of most established tissue cell cultures.

Synthetic of chemically defined growth media have also been developed. Their Exact composition are Known, and can duplicated as needed. Generally such synthetic media are complex, containing a rich variety of organic ingredients including amino acids, carbohydrates coenzymes, lipids and vitamins. The particular concentrations and combinations of these substances are determined by extensive studies, the results of which yield an exact and suitable growth medium. It is completely defined concoctions.

Aspect prepautions are exercised in preparing various types of media used in tissue cultured procedure. Balanced salt solution may be sterilized by autoclaving. However, organic components of media, such as sera and amino acids are sterilized by filtration, passage through asbestos (Seitz) or Millipore filters. Many investigators prefer to filter all solutions and then combine the various media components in desired proportions. In order to keep tissue cultures free from unwanted bacteria and fungi, antibiotics are commonly added.

Representative Technique Culturing Cell

The source of material for the majority of primary cultures can be divide into adult, embryonic, and tumor tissues. Unfortunately, many variety of cells derived from such tissues grow poorly or not at all in vitro. This obviously limits the type of cell that can be studied and used for the cultivation of viruses. Adult cell culture most commonly sued are derived from the foreskin, the kidney, the lung, and the thyroid. Cultures of adults kidney from every available animal have been quite successfully and widely sued. Embryonic cell cultures have been obtained from variety io animals, including chicks, mice, and humans. Several human carcinomas (malignant growths composed of epithelial cells) have served as source of many continuous cell lines.

The cell used in most studies are of to general forms: cell sheets (*monolayers*, or single layers of cells), and suspensions of tissue fragments or dispersed cells.

Cell sheets

The initial cultivation if cells from tissue is commonly referred to as the *primary culture*, in contrast to subculture or transfer. Primary culture have certain disadvantages. Including the presence of a variety of cell types and the possible presence of contaminating viruses. The development of pure cultures of cells can pose several problems, but such preparations are possible to obtain. When cells of these generations (usually measurable in yeas), this systems are called *continuous* or *established cell lines*. It is important to note that pure cultures originally develop from isolated cells. The subsequent cells are descendant of single cell an constitute a pure culture, called a *clone*.

The U.S. Department of Health , Education, and Welfare issues the publication "Registry of Animal Cell lines certified by the cell culture collection committee" which contains complete description of cell lines used in vaccine production and scientific investigations.

The monolayer procedure includes treating minced tissue enzymatically with trypsin for period of time. Freed cells are washed, diluted in appropriate quantities, suspended in a suitable growth medium, and seeded, that is, dispensed into any of a variety of tissue culture vessels including Petri dishes, bottles, culture flasks, and screw-capped tubes. Most primary cells can be subcultured to produced secondary cultures. Morphology, subcultured cells are similar to primary cells. Cells of this type comprise *cell-strains* and apparently can be subcultured indefinitely.

Suspensions of tissue fragments and dispersed cells : suspended cell cultures

Monolayers have largely displaced cell suspensions, but in particular situations the use of dispersed cell cultures has certain advantages. Techniques employing such systems have been used for large-scale production of viruses and of mammalian cells.

In tissue fragment procedures, finely chopped pieces of kidney or other tissue are suspended and maintained in an appropriate medium. The original technique using this type of system was that of the Maitlands, mentioned earlier.

Dispersed cell suspensions simply consist of cells submerged in a nutrient fluid. In some cases, the cells are stirred continuously to prevent them from sticking onto the sides if the tissue culture vessel.

Preparations of this type are also used to obtain large quantities of cells. Dispersed cell suspensions can also be employed to detect antibody activity against viral pathogens. Normally, in a suitable growth medium cells produce acidic waste products which lower the pH of the system. This change in the pH is shown by an indicator such as phenol red, which in alkaline environments is red, but in acidic states is yellow. When a virus inoculum is added to a dispersed cell suspension, the cells of the system are prevented from metabolizing normally and the colour of the medium remains red. However, if specific antibodies against the viral agent (from the serum of a test animal) are introduced into the system at the same time as the virus, in a "serum-virus mixture," the cells metabolize normally. The medium will turn yellow, thus showing the neutralizing capability of the antibodies.

Organ cultures

Fragments of tissue may also be cultured in plasma clots. Preparations of this type were among the earliest tissue cultures. The technique basically consists of placing a piece of tissue in a small drop of sterile plasma which is on a glass surface of some type, such as the inner surface of a test tube or a cover slip within a glass vessel. The preparation coagulates quickly. In some procedures, after coagulation takes place additional nutrients are added in the form of a liquid medium. The container is stoppered, and incubated at approximately 37°C. An outgrowth of cells occurs within a few days.

Several variations of the original procedure developed by Alexis Carrel (Nobel Prize laureate, 1912) are known. One of these is the so-called *organ culture technique*. This procedure has been used in studies concerned with the development and growth of various human respiratory viral pathogens. The original techniques of organ culture were devised by Fells and Robinson. Tissue fragments, *explants*, were implanted onto plasma clots which contained chick embryo extract diluted with BSS and glucose. The tissue was transferred to new clots each day. A raft of rayon, suitably treated so that it would not be toxic, was introduced later to serve as a support for the explants. Modifications of this technique have appeared using a raft in a Petri dish and adding a liquid medium to provide the essential nutrients for the tissue fragments.

Detection of Viral Multiplication in Tissue Culture Systems

Several tissue culture indications of viral infection have already been described, including cytopathic effects and alterations in cellular metabolism. As this general subject is an important one, certain features of cytopathic changes following cell infection and certain techniques used in the enumeration of viruses will be described here in some detail.

Cytopathic changes

Gross cytopathic changes in virus-infected cultures can be readily observed with the aid of standard light microscopy techniques. Although fixation and staining of infected cells need not be carried out for detecting virus multiplication, this procedure is used for keeping permanent records of its stages.

Characteristic changes brought about by several viruses include cytoplasmic granulation or vacuolation, formation of giant and multinucleated cells, and condensation of chromatin material in the nuclear membrane of cells. The particular type of CPE (cytopathic effect) produced can at times be used as an aid in diagnostic virology, the identification of a particular viral group. Certain viruses, however, characteristically do not cause any observable morphological change in cells.

It should be kept in mind that newly isolated viruses occasionally must be transferred several times before the characteristic CPE are produced. It is also important to note that several changes in the fine structures of cells may be the result of the effects of a temporary lack of essential nutrients, changes in pH, and other related factors. Generally these effects develop within 24 hours and are not neutralized by antiviral substances, such as antibodies.

Plaquing techniques

Dulbecco, in 1952, introduced the plaquing method for viruses. The procedure consists of first inoculating a monolayer tissue culture system with a viral suspension and allowing sufficient time for the viruses particles to be absorbed by the tissue cells. The system then is overlaid with the nutrient agar and incubated at the desired temperature. Viruses that infect the cells multiply within their hosts and produce cellular damage. The presence of the agar overlay limits the spread of these agents to a lateral direction, producing circumscribed areas of cellular degeneration. Such sites of destruction, called plaques, are detected by staining either the remaining living cells, while trypan blue can be employed for dead cells. The characteristics of the plaques formed (e.g., diameter and margin properties) in a particular type of tissue culture system such as monkey kidney or human kidney or HeLa cell also helps in the identification of viruses.

The plaque technique can be used for the establishment of pure viral lines. This procedure, called *plaque purification*, is based on the presumption that a single virus Particle produces each plaque.

Plaquing methods can be carried out in a variety of bottles, Petri dishes, and related vessels. These procedures are used in *plaque assay* to determine the number of infectious virus particles in a particular specimen. The measurement of infectivity is accomplished by carrying out the plaque technique with given dilutions of virus-containing suspensions. The infectivity of the original virus-containing suspension is computed from the number of resultant plaques, each of which is referred to as a *plaque forming unit* (PFU). Together with a correction factor for the dilution of the dilution of the specimen employed. Modifications of this technique are necessary at times, and are determined by the type of virus being assayed.

Cultivation of Plant Viruses

Dr. P.R. White (1934) was the first to grow plant viruses in tissue culture. He could successfully cultivate tobacco mosaic virus in excised tomato root tips.

The method he developed was simple. It has since been widely practised by plant virologists all over the world. In this method, a systematically virus infected tomato plant was used. Its stem was cut up into segments which were washed thoroughly and were suspended by threads in conical flasks containing little water. Care was taken to see that the segments touched neither the water nor the side walls. The segments were allowed to develop roots. After a period of time, the root tips were removed and placed in flasks containing 50 ml of a suitable nutrient medium.

Chemical composition of one such medium is as follows :

$Ca(NO_3)_2$	0.60 millimols
$Mg\ SO_4$	0.30 millimols
KNO_3	0.80 millimols
KCL	0.87 millimols
KH_2PO_4	0.09 millimols
$Fe_2(SO_4)_3$	0.006 millimols
Surase	2 per cent
Yeast and extract	0.01 per cent

The tips were allowed to grow for about a week. Afterwards, the surviving cultures were cut into uniform length of about 1 cm. These were then subcultured in fresh medium as before. The root tips of the subcultures were removed and one or two of these were selected and stocked for further subculturing. Viruses were found to be multiplying in these tips and could be sued for studies taken up subsequently.

Using the above technique it has been possible to culture such viruses as tobacco necrosis virus and wound tumour virus. The degree of success achieved with different viruses has been variable. It has been found that the type of virus, the nature of the host and nutritional factors considerably affect the growth of viruses.

Plant viruses are normally cultivated on tissues obtained from already and systematically infected plants. This is because it has been found to be difficult to infect healthy plant tissue cultures freshly with a virus.

Nevertheless, the need for *de novo* culturing of viruses sometimes arises. In such cases healthy tissues, not necessarily of the natural host, are first cultured. Later, it is infected artificially by the virus to be grown. Methods to do so have been developed and perfected in recent years. In some cases the viral inoculum is directly poured on the growing tissue. To make the infection steady and quick, the tissue surface is pricked with the needle or rubbed with abrasive quartz or carborandum powder. Sometimes, the natural vector of a disease, which itself harbours the virus particles, is used to inject the cultured tissues.

A more recent development has been the use of isolated plant protoplasts for culturing plant viruses. This method has the additional advantage in that protoplasts can be uniformly and synchronously cultivated.

Isolation and Purification of Viruses

We have seen how viruses are cultivated. However, in order to study them thoroughly one must have them free from unnecessary contaminating material, such as components of the tissue cells. It is essential that the viruses are available in a reasonably pure form. To achieve this they must be isolated first and then suitable purified. The principal techniques used for this purpose are (a) ultrafiltration; (b) ultracentrifugation; and (c) chemical precipitation. In recent years chromatographic, electrophoretic and solvent extraction techniques have also been increasingly used.

Certain points should be given the proper attention before any of the steps for isolation and purification can be taken up. These concern the availability of virus particles in active, infective form in the initial extract. Viruses, which are totally host dependent, remain, while inside a cell, sort of insulated against the vagaries of changing external temperature and pH conditions. Therefore, firstly, the operations for isolation have to be carried out at extremely low and constant temperatures. Thus adverse effects of temperature changes can be minimized, if not completely removed.

Secondly, the selection of proper pH conditions of the extracting medium is important. Suitable buffers have to be used. Both inorganic and organic buffers are in use. With experience the choice of buffer for a particular type of virus is standardized to ensure minimal loss.

Ultrafiltration

This operation is based upon the fact that, generally speaking, virus particles have a smaller size dimension than any other microorganism. Therefore, to separate viruses from amongst a mixture of variously sized particles, one has simply to use a filtration device capable of retaining the larger particles while allowing the smaller virus particles to pass through. For instance, let us suppose a viral suspension present in a bacterial medium is to be isolated by filtration. All one has to do is to allow the fluid suspension to pass through a suitable filter capable of allowing only the virus particles to pass through, whole keeping back the larger particles. Seitz type filters or porcelain G-5 filters may be conveniently used. The clear filtrate thus obtained should contain solely viruses.

In order to retain these virus particles the viral suspension may further be passed through ultrafilters made of cellulose acetate or collodion membranes. The very small pore size of these filters enables them to retain the comparatively larger particles of the virus. Such membrane filters with graded porosity, like Millipore filters (manufactured by the Millipore Filters Corporation, USA) have accurately determined pore size. These can, therefore, be used to segregate particles having different sizes. Particles larger than a particular pore size get wedged in the small pores of these filters. Thus the smaller ones get through but may again be retained by filters having still smaller pore size. The filtered particles of a particular size cab be brought in fluid suspension by washing the filters in reverse direction. In recent years, diatomaceous earth has been increasing in use as a filtering aid. Charcoal and bentonite have also been successfully used. More sophisticated gel filtration techniques using agar-agar or 'sephadex' allow selective filtration of virus particles.

The filtration of viruses, no matter how small be the pore size, is not fool-proof against possible presence of other organisms. This is because there are a few other types of microorganisms which have size ranges comparable to that of larger viruses. The Rickettsiae, a pathogenic group of micro-organisms and mycoplasmatales, a type of bacteria are two examples of such organisms. Therefore, the technique of ultrafiltration is likely to be useless in case the viral suspension contains these or similar micro-organisms as well.

Ultra-Centrifugation

Out of the particles in a suspension the heavier ones have a tendency to settle down first. This is because of the attraction of gravitational forces. Usually, the process would be a slow one. In case of extremely small particles, the process of settling down or *sedimentation* may take a almost indefinite period. The rate of sedimentation could be increased by increasing the gravitational pull itself. This can be done by applying centrifugal force on the suspension. This is commonly achieved by rotating centrifuge tube containing the suspension in a centrifuge machine. Such rotation or centrifugation at a very high speed produces forces equivalent of many thousand times of the gravitational force. Depending upon the mass of the particles to be segregated heavy particles like cell debris, centrifugation at relatively low speeds are enough. However, when lighter particles are to be sedimented, centrifugation at high speed becomes essential. Centrifugation at extremely high speed, generating thousands of times of normal gravitational pull, is known as *ultra-centrifugation*. The machines designed to achieved this are called ultra-centrifuges.

Virus particles present in a suspension containing cell residues and other materials may be a purified by a two step centrifugation process. In the first step the residual cell material or bacterial cell *et cetera* are separated out by centrifuging the suspension at low speed. The semi-purified suspension could be further purified by filtration. The filtrate containing the virus particles may then be subjected to ultra-centrifugation. The virus particles sediment and form a small heap or pellet at the bottom. This technique is called differential centrifugation. In this manner Schlesinger in 1933 was able to isolate the bacteriophage *WLL*.

In recent years, the technique of ultra-centrifugation has been modified to ensure greater solution. Two commonly used modified techniques are known as rate-zonal centrifugation and equilibrium density gradient centrifugation. In these methods, the viral suspension is mixed with appropriate gradient concentrations of sucrose and on caesium chloride in sodium chloride solution. The gradient is previously prepared by successively pouring equal volumes of sucrose of caesium chloride

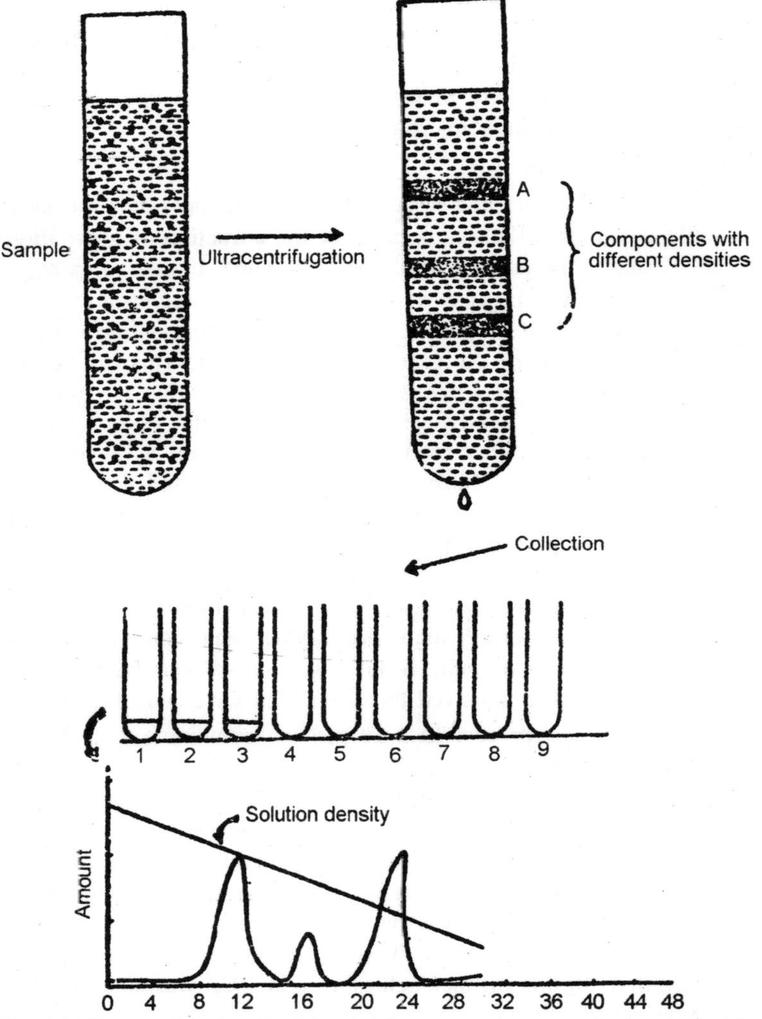

Fig. 6.1. Ultracentrifugation. Particles of different mass size present in the same medium cn be segregated by this method.

of decreasing concentration into .14 M sodium chloride solution taken in plastic centrifuge tubes. The mixture is then allowed to equilibriate at 4°C for 12 hours for achieving a stable gradient.

The mixture is subsequently centrifuged at a very high speed (40,000 × g – 150,000 × g). In rate zonal centrifugation the duration of centrifugation it could be for several hours or even days. On centrifugation in this manner, the virus particles do not settle down. Rather they aggregate at certain regions in the centrifugate, forming distinct bands or layers.

The number of such bands indicates the number of types of particle in the suspension. In rate zonal centrifugation, the particles separate out and form bands relative to one another, on the basis of their relative densities. The bands formed are not stable and would normally be displaced on further centrifugation. However, by this method, segregation of the components in a mixture is easily achieved.

In equilibrium density gradient centrifugation the components of identical densities form bands with stabilized position in the centrifuged fluid. This is because the bands are formed in the regions of the solution with densities in equilibrium with that of the virus particles assembled at that region. These bands would not be displaced even on prolonged centrifugation. This method is, therefore, also known as *isopycnotic* density gradient centrifugation. The different layers may be observed by placing the centrifuge tubes against suitable optical systems. These layers can then be separately collected by punching the tubes at the respective position and by drawing out the fluid cautiously by a syringe. Alternatively, the liquid in the tube could be allowed to flow out and collected drop by punching the tube at the bottom. Utilizing rate zonal-centrifugation technique it has been possible to purify Shope papilloma virus and poliovirus. Characterization of virus particles using equilibrium density gradient centrifugation have been achieved in case of $\phi \times 174$ and tobacco mosaic virus strains. Some animal viruses have also been purified utilizing this technique.

Precipitation

The most effective method for obtaining viruses in the pure form is crystallization by chemical or by physical means. The suspension or extract containing virus particles is initially purified using the techniques of ultrafiltration and ultracentrifugation. The purified preparation containing the virus particles is then precipitated. Chemical agents like ethanol or ammonium sulphate has been found to be the effective concentration for precipitating viruses. The precipitates obtained once could be redissolved in some suitable buffer solution and reprecipitated. In this manner more and more purified preparations are made available. Methanol, protamine and even yeast nucleic acid have all been found to be effective precipitants of animal viruses.

Precipitation can also be achieved by adjusting the hydrogen ion concentration (pH) of the suspension. Proteins, as you know, are ampholytes i.e., they possess two types of surface charges, positive and negative, on them. Proteins remain the solution till the charges are unequal in number and till they are all neutralized. Once that happens they precipitate out. For a particular type of protein only one pH level is capable of precipitating that type of protein. This pH is called the *isoelectric point*. pH can be adjusted by adding more cation (Mg^{2+} Ca^{2+}, H^+ etc.) and anions (Cl^-, PO_4^{3-}) in the form of salt solution. These ions neutralize the negative and positive charges respectively and the isoelectric point is reached. Thus proteins are precipitated out. Since viruses have an outer protein coat, they can also be precipitated in this manner. Alfalfa mosaic virus is one of the first viruses to be precipitated by this method.

Absorption Chromatography

Apart from these three major techniques, a few more effective physico-chemical methods have been developed and used in recent years. One of these is absorption chromatography. This is dependent upon the surface properties of the protein coats of the virus particles. Because if these the virus can get associated with or absorbed on the surface of a solid but porous substances having similar properties (charge, etc.) on them. Substances like calcium or aluminium phosphate are generally used. These are packed on columns of glass tube designed specially to hold them. The packing of the chromatographic material has to be done under special conditions.

The viral suspension along with impurities is then poured on the column. The liquid is allowed to drain off by keeping the outlet stopcock of the column open. Virus particles get absorbed onto the surface of the material. Water flows out and with it some of the impurities. The absorbed virus particles can be reclaimed by washing the column with a large volume of specific medium such as $MgCl_2$ solution of an ionic concentration higher than that of the viral suspension. The virus particles come our because they have greater affinity for the washing medium. The reclaimed virus is purer than it was earlier. The presence of virus could subsequently be detected in the outflowing fluid (eluate) which is collected in small volumes. Ion exchange resin like, Dowex I have also been successfully used. Separation on columns of neutral substances like silica gel. Calcium phosphate and magnesium pyrophosphate have also been reported.

Solvent Extraction

Purification of viral preparations may also be achieved by what is known as partition between solvents. It is well known that a substance

Fig. 6.2. A diagrammatic comparison between the resolution of image of an object by electron microscopy and by light microscopy.

could be soluble in more than one solvent. Usually the solubility of the material in these solvents (the amount that can be dissolved in a fixed volume of the solvent) also differs. In other words, a substance is more soluble in one solvent than in the other. If we wish to purify a virus preparation containing some impurities, then the suspension is first mixed with one of the solvents. Since the suspension is in water, it forms one of the solvents. The second solvent has to be immiscible with water.

The entire mixture is then shaken vigorously and allowed to stand. If the impurities are more soluble in the second solvent than they are in water, they gradually pass into it. Virus particles remain in the water medium. In this manner, impurities are eventually removed and the viral preparation becomes purer and purer. Since the two solvents are immiscible, the separation of pure suspension and the impurities is almost total. Organic solvents like fluorocarbon and dextran are often used as the second solvent.

Other Techniques

Apart from the above mentioned techniques, viruses have been reported to be successfully purified using serological and electrophoretic principles. However, these methods are of only selective and limited utility. Certain impurities associated with various preparations, particularly from plant material, have been found quite difficult to get rid of, with the help of techniques discussed above. For instance, such impurities as protein pigment complex or cellular nucleic acids often prove bothersome. Enzymatic treatment of impure preparations have been found to be useful in such cases. Use of proteases and nucleases is quite a routine these days.

All these methods and other common laboratory techniques are used together to isolate a virus from its natural source. Depending upon the source, the nature of impurities and the probable concentration of the virus particles in the source, the nature of operation differs. Various procedures have been developed in different laboratories in various parts of the world to isolate and purify virus samples from plant and animal sources. We shall consider here briefly the basic features of these procedures as applied to different sources.

Isolation and Purification of Plant Viruses

Modified after Markham and Smith 1949

The infected plant materials or tissue and cell cultures harbouring viruses particles are first collected. The infected tissues are then cut into small pieces and then homogenized by any available means. A suitable buffer is used as the extracting medium. The extract obtained is then filtered through a piece of fine cloth to separate the cell debris. The filtrate is then centrifuged at low speed (2500×g) to remove the remaining cell residues.

The crude preparation obtained in this manner contains soluble impurities like plant proteins. Gummy substances may also be present. Plant proteins could be selectively precipitated at this stage by heating crude preparation upto 55°C. Most viruses are found to remain stable even at this temperature. However, some may get destroyed. This treatment, therefore, should be applied with caution. Selective precipitation of the bulk plant protein could also be achieved by treating the suspension with a specified amount of ethanol.

Proteins can also be removed by treating with chloroform and centrifuging the mixture at slow speed (2500×g) for 20 min. Proteins mostly go into chloroform layer or are aggregated at the interface and the aqueous layer contains viruses.

The semi-purified preparation thus obtained can be further purified by precipitation of the virus particles with 30 per cent ammonium sulphate or with cold ethanol. Saturated ammonium sulphate solution is added to the suspension till the final concentration reaches 30 per cent level. Proteinaceous virus particles are precipitated out along with other protein impurities that may still be present in the original suspension. These are then collected by centrifugation. If desired, the precipitate thus obtained may be resuspended in buffer and reprecipitated in the same manner. This step could be repeated a number of times. At each step the purity of the preparation is found to be more than it was earlier. The pellet obtained ultimately may be dialysed against a suitable buffer to remove excess of ions. The dialysate is then centrifuged at $60000 \times g$ and the pellet obtained resuspended in the same buffer.

The considerably purified virus preparation is then further purified by absorbing the particles on a suitable chromatographic material in a column like that of calcium phosphate or of Dowex I, and then reclaiming them by means of elutant or washing solution. Most of the impurities present till now get removed by this technique. Impurities that cannot be absorbed by the column simply move out with the outgoing buffer solution. Protein impurities, of course, get absorbed on the column. However, very little portion of it is likely to come out with the washing solution. This is because of the low affinity of these proteins towards the washing solution.

The virus particles present in the washing solution may then be passed through a membrane filter capable of retaining viruses. Such retained particles can easily be resuspended in a small volume of the buffer. This suspension may again be reprecipitated with ammonium sulphate. The precipitate collected by centrifugation has considerably purified virus particles. Repeated ad careful precipitation in cold temperature conditions are likely to yield crystalline preparation. In fact, to ensure that the physical characteristics of the viral preparation remain little disturbed, the entire operation should preferably be carried out at low temperature conditions. These may further be purified by density gradient ultra-centrifugation.

It should be remembered that the same process is not equally effective in the isolation of plant viruses from different sources. In recent years several modified procedures have been developed to meet challenges offered by different situations.

Preparation of Animal Viruses

Modified after Crammer 1963

Isolation and purification of animal viruses is more complicated. Infected parts of an animal or tissues cultivating animal viruses are first collected. These are then homogenized in a cold with a suitable buffer. On proper homogenization, the tissues get completely disrupted and an uniform suspension is obtained. This suspension is then centrifuged, first at slow speed (1400×g) and then at high speed (7000×g) for 20 minutes duration each time. Impurities sediment down and a clear suspension is obtained.

The suspension is then passed through a bacterial filter to get rid of most of the larger suspended particles. The filtrate is then passed through a membrane filter to retain the virus particles. The filtered particles are then resuspended in a small volume of a suitable buffer. The virus preparation at this stage contains some cell proteins and liquids as impurity.

This suspension is then ultra-centrifuged at a very high speed (20,000×g) for several hours. Virus particles sediment out and a pellet is formed at the bottom of the tube. In certain cases ultra-centrifugation should be done with caution as it might disturb the physical architecture of the viruses. The centrifuged virus particles are then suspended in a small volume of the buffer. At this stage it could be stored for further purification at a later date. For this purpose, the suspension may be evaporated to dryness in vacuum or dried under freezing conditions (lyophilization).

At this stage, the partially purified virus preparation contains a reduced amount of both proteinaceous and non-proteinaceous impurities. The latter, like lipids and carbohydrates, are separated first. Lipids can be removed by treating the preparation with some lipid dissolving but harmless (for virus) organic solvent. Since these solvents are immiscible with the buffer used, the dissolved lipids are all present in the organic solvent. Repetition of this process makes the preparation almost lipid free. Polysaccharides, if present, may be removed at this stage by electrophoresis. It would also be removed by non-exchange chromatography at a later stage.

The sample thus becomes ready for chemical precipitation. This is done by treating it with ethanol or with 30 percent ammonium sulphate, final concentration, for 12 hours or more. Precipitation can also be achieved by adjusting the pH of the suspension to the isoelectric point of the viral proteins. This is preferred in certain cases where protein impurities are still there and can only be removed by selective precipitation.

The precipitate is then collected by centrifugation at high speed and resuspended in a small volume of the buffer. It is then dialysed against a suitable buffer. It may be further purified by absorption chromatography as detailed earlier. Density gradient centrifugation is useful at this stage; virus particles of the same molecular weight may thus be segregated. The preparation at this stage is almost pure chemically. Crystallization may be attempted now. The particles are resuspended as usual, and precipitated under controlled conditions. Precipitates formed gradually crystallize out.

The general procedure outlined above can be used for isolating and purifying viruses from other sources, like insects or bacterial lysates, as well. Individual cases, of course, necessitate modifications.

Criteria of Purity of a Preparation

When a sample has been purified in this manner the question that comes up is about its purity. This question has two aspects. The purity or the homogeneity status has to bee considered from the (i) physicochemical as well as from the (ii) biological point of view.

Chemically, a preparation may be regarded as the purest when it is available in the crystallized state. Even in suspension, an extremely purified sample shows certain characteristics of its own. For instance, a purified suspension of tobacco mossaic virus shows a very brilliant shine. In most cases an unusual brightness is visible. Physical criteria like size, shape, solubility and molecular weight are also helpful in establishing the purity status. In recent years, electrophoretic and serological data have been found useful in determining the homogeneity of a sample. However, even in a state of high apparent purity, foreign material like protein molecules of the host may still be present. Absolute purity is, therefore, rarely, if at all, achieved.

Biological purity of virus preparations has also to be ensured. Biological purity relates to the capacity of the preparations to infect a host. If this capacity is lost during the preparation of the sample, it would biologically be impure. This capacity should, therefore, be tested at every step to see it is still there. With successive stages of purification, the infectivity tends to get diminished with relation to number of observed particles. If proper precaution is maintained throughout the operation, this loss can be substantially checked. However, absolute constancy in relation between infectivity and umber of particles at every range is rarely achieved.

Cultivation and Enumeration of Bacteriophages (Phages)

Numerous types of bacterial viruses are known each capable of multiplying within one or more different bacterial species. In general, the cultivation of these parasitic forms is relatively uncomplicated, neither expensive nor overly time-consuming. For these reasons, bacteriophages have been used by many virologists to study the mechanisms of virus infections and related phenomena. Phages are commonly propagated on appropriate, actively growing young bacterial cells. Either broth or agar cultures are used for this purpose. In liquid cultures, the lysis of sufficient numbers of susceptible bacteria will cause the nutrient medium to clear. When agar plate are used, the presence of bacterial viruses is indicated by the appearance of transparent holes (plaques) against the dense background of bacterial growth on the medium's surface.

Various types of bacterial viruses can be isolated from natural sources, such as dairy products, diseased tissues, feces, sewage, soil, and water. The demonstration of such viruses and their bacteriolytic action may be accomplished by passing a sufficient quantity of fluid samples of any of the aforementioned materials through a filter that retains bacteria, but not viruses. Examples of such devices include the Seitz (asbestos) and Berkefeld-N-Type filters. The next step of the procedure determines whether the resultant filtrate material contains phage. This generally involves introducing a small account, such as 1 ml, of the test material into approximately 10 ml of an appropriate fresh culture of a test host organism. Bacteria in the log phase of growth are suitable hosts. Another tube, without the addition of the suspected phage-containing sample, is generally used as a control. Both tubes are then incubated under optimum environmental conditions. If the tube to which the test material was added shows clearing or at least appears to be less turbid than the control, phages were present in the original sample. Unfortunately, the demonstration of bacterial viruses from natural sources is not always so easy. In many studies, additional and more involved procedures are required. Moreover, many samples of natural substances may not yield bacterial viruses.

Once a new phage is found, its characterization is usually undertaken. This involves determining its chemical and physical properties, uncovering its mechanism of bacteriolytic action, and demonstrating its basic differences from and similarities to other bacterial viruses.

Enumeration of Phages

Several methods may be used to estimate the number of virus particles in a sample accurately. One of the most useful and commonly employed procedures is the plaque count assay, which is simple, accurate, and yields highly reproducible results. It was originally introduced by D'Herelle, a French Canadian, around 1917, and was perfected by Gratia in 1937. It was D'Herelle who coined the term bacteriophage from the Greek word *phagein*, meaning "to eat".

The procedure can be performed by adding an adequate quantity of a given phage suspension dilution to a tube containing 2.0 ml of melted soft agar (SA) kept at 45°C. One drop of an appropriate fresh bacterial culture is also introduced. The contents of the tube are mixed and quickly poured onto the surface of a dried hard agar (HA) layer contained in a Petri dish. This plate then is rocked gently back and forth in order to equally distribute both the bacteria and the phage over the agar surface before the SA hardens. Viral particles diffuse through the medium, infect, multiply, and lyse susceptible bacteria. Those bacterial cells not infected grow, multiply, and form a "turbid" layer of film over the hard agar surface. Evidence of viral infection is provided by the presence of the previously mentioned plaques. As in the case of the plaque assay for animal viruses, the number of viral particles in the original sample is determined by the number of plaques multiplied by the dilution factor, the reciprocal of the dilution. This value is expressed as the number of plaque-forming units (PFU) per milliliter of the initial sample.

The quantitative determination of virus particles contained in a particular specimen can also be made by electron microscopy.

Phage Typing

Bacterial viruses that are continuously propagated in a specific bacterial species can become adapted to that particular bacterial host. Such viruses exhibit the phenomenon called *host-controlled modification*. They will primarily infect and lyse a specific bacterial species or cells of associated strains. A practical application of this process, namely phage typing (Colour photograph 54), is used to identify strains within certain bacterial species, such as *Salmonella typhi* and *Staphylococcus aureus*. This procedure is an extremely important tool I tracing the sources of epidemics and distinguishing between pathogenic bacterial strains which cannot be differentiated by other means.

Phage typing is performed by growing an unknown bacterial culture-isolated from a patient, for example-on an agar plate. Specific phages are systematically spotted onto the "lawn" of bacteria. After a suitable incubation period, zones of lysis appear if the appropriate phage and bacterial host combination is present. Plastic plates with grid patterns embossed on their bottom surface are available to simplify the procedure.

Enumeration of Viruses

Enumeration of virus particles in a suspension is considered very important. This is because the knowledge about the number of particles present in a suspension becomes a prerequisite for further studies. Information regarding the relationship between infectivity and number of particles is also significant.

Counting by Direct Electron Microscopy

Often counting is simply done by direct observation under electron microscope. One widely used procedure is called *Latex droplet method*. In this method a known volume of the suspension is spread on a cellophane film after being mixed with known number of polysterene latex droplets. The film is dried and observed under the microscope. Suitable techniques like metal shadowing, increases the resolution of the particles.

Virus particles only get coated with the latex and are seen very clearly. Free latex particles, which are comparatively larger in size are also visible side by side. Both the types of particles are counted and their relative proportions known. Since the concentration of the latex particles sprayed as sell as the original concentration of the latex particles per ml. is pre-determined, it is easy to calculate the total concentration of virus particles.

Plaque Assay

Another method of enumerating viruses is the Plaque Assay. In this method a suspension culture of appropriate host cells is prepared initially and later cultured in petri plates containing suitable agar based media. Such cultures are then inoculated with aliquots of the virus titre to be enumerated. On proper incubation virus particles start infecting the host cells. One susceptible host cells is infected by a single virion. After some time, disintegration of the host cells occur resulting in liberation of more virus particles. These, in turn, infect host cells in the immediate vicinity and destroy them as well. In the meanwhile, the cells that are not infected continue to grow and divide normally. Thus, soon transparent, visible, circular areas appear on the beneficial lawn in the plate. These are called plaques.

Since each virion of the original suspension causes one plaque, the number of virus particles in it can be determined with ease. For example, let us suppose 1 ml. Of the original suspension was diluted 10^4 times before being distributed with the host cell culture; and let us say 42 plaques were formed. Obviously, there were 42×10^4 particles per ml. In the stock. In this method, virus suspensions are often mixed with the host cells prior to incubation. Plaque assay has been most effective in the counting of bacterial viruses.

Acid End Point Method

It is well known that vigorously growing tissue or host cell suspension turns acidic the medium in which they are growing. This feature is taken advantage of in this method. It involves measurement of any change in the acidity of the suspension and correlation of the data with standards prepared earlier. Here, the idea is to determine the highest dilution of a suspension which would be capable of causing infection. Thus a medium which is not acidic to the normal level would indicate the presence of deed cells. When virus particles are cultivated with the tissue suspensions, death could be due to viral infection alone.

Supposing we wish to find out the number of infectious virus particles present in a particular suspension. Initially, it is diluted, let us say, 10^4 and 10^5 times, separately. Both the dilutions are then made to infect identical host cell suspensions. Let us assume that with the former no discernible acidity developed whereas with the latter the medium became acidic developed whereas with the latter the medium became acidic to the normal level. This shows that when the original suspension of virus particles was diluted 10^5 times there was not even a single virion in it. The more concentrated 10^4 dilution, on the other hand, contained at least one virion. Therefore, the original suspension had between 10^4 and 10^5 virions. There are several procedural modification of this procedure which are in use in various laboratories. Two common procedures are the metabolic inhibition test and the 50 per cent end point test. These methods are not, however, very precise and need many replications to arrive at a satisfactory result.

Haemagglutinin Assay

For containing viruses in the *Haemagglutinin assay*. Many viruses are capable of agglutinating red blood cells. This phenomenon is known as agglutination or sticking together of red blood cells, and the agent responsible, Haemagglutinin. A virus particles attaches simultaneously to two red blood cells and bridges them. Such formation aggregates can be detected in a number of ways and afford a simple method for virus enumeration.

One of the simplest and most commonly use haemagglutinin assay procedure is the *pattern method*. This is based on the observation that when the suspension of blood cells and the virus suspension is left undisturbed in a small tube for several hours, non-aggregated and aggregated cells behave differently. Whole the former sediment to the bottom of the tube and then roll toward the center, the latter only sediment to the round bottom but do not roll. The aggregated particles rather form a thin film that is very characteristic of the particular virus.

The actual assay is carried out by an end point procedure. Serial dilutions of the virus sample as taken in replicates and mixed with a standard dilution of red blood cells (usually 10^7 per ml). The last dilution showing complete hemagglutination is regarded as the end point. This value could be regarded as en estimate of the viral population. This value could be regarded as an estimate of the viral population. The titre estimated by this method has, however, an inherent imprecision at least as large as the dilution step used (usually two fold). Colometric measurement of the experimental titres and their calibration on a graph gives more precise information.

Particle Count and Infectivity

Studies on the enumeration of viruses reveal an interesting situation. It has been found that in many cases there is no one to one correspondence between the number of particles as physical entities and particles capable of infecting a cell. In the case of tobacco mosaic virus the frequency of occurrence of an infective particles may be 10^6. While some of the methods described above tell us about the number of particles others indicate the number infective units under the given conditions. The data obtained should, therefore, be assessed with caution. It is, of course, always possible that even those particles which are non-infective under the given conditions are fundamentally capable of successful infection.

Serology

Serological techniques are also in extensive, though restrictive, use in the study of viruses. It is common knowledge that when an animal (vertebrate) is infected with a foreign material, be it a virus or bacterium or a macromolecule like proteinst there are produce in its blood stream proteins which combine specifically with the substance entered. This act of specific combination can be demonstrated *in vitro* and forms the basis of serological tests.

The proteins specifically produced in the blood stream as a result of external stimulus are called Antibodies. Any substance which will stimulate the production of antibodies in vitro is called an *Antigen*.

Blood serum containing antibodies is termed *Antiserum*. The serum obtained from an animal not injected with a specific antigen is called Normal serum. Antigen-antibody reactions have been used in serological studies of viruses in several ways. These are :

Neutralization reactions

Properties, particularly that of infection, of a virus acting as an antigen, are neutralized.

Complement fixation test

When antigens are mixed with their specific antibodies, the mixture has the property of removal of the power of the normal serum to haemolyze red corpuscles. Complement is a neat labib protein present in normal blood serum and vitally necessary for some antigen-antibody reactions to occur at all. Complement when 'fixed' by antibodies do not allow normal haemolytic reactions to occur. This can thus be helpful in detecting antibodies in unknown serum.

Precipitin reaction

When a virus comes in contact with its specific antiserum taken in saline at different dilutions, and warmed in a water-bath, a precipitate is formed. The antibody is referred to as the precipitin and the reaction as precipitin test.

Anaplylaxis

In this type of test, the union between antigen and antibody is detected by reaction offered by animal tissues. It is an *in vivo* system of detection of antigen-antibody reaction. Out of these, the precipitin reaction and complement fixation test are extensively used in virological studies. Additionally, this property has been exhibited in the new universal process of artificial immunization.

Precipitin Reaction

To proceed with this reaction purified antigen (virus) has to be prepared first. When virus antigen is being prepared from tissue extreme care has to be taken to eliminate any possible protein or similar contaminant which themselves might act as antigen and confuse the results. There are instances when crimeous conclusions have been drawn only because the antigen preparation were contaminated. Usual procedures adopted for preparation of purified samples of virus have been outlined earlier. Antiserum is obtained by injecting the purified antigen (virus) into a suitable animal like rabbit. Domestic fowl and also the horse are sometimes employed. The antigen is usually injected into the vein which runs along the upper surface of the ear. About two weeks after injection (the incubation period) the rabbit is bled from the other ear. A small cut is made near the base of the ear. The blood is collected in the tube and is allowed to clot. The serum is decarted off and centrifuged to remove any remaining blood cells. Antiserum prepared in this way is not sterile and has to be stored with proper precaution to prevent bacterial growth.

In the earlier days of serology, the precipitin test was carried out in liquid medium. A convenient method was to use 1 ml. Samples of antiserum at constant dilution in a series of test tubes to each of which 1 ml. Of antigen solution at different dilutions was added. The tubes were then set after a brief shaking in a water bath with temperature at 50°C. The tubes were then observed for the appearance of precipitate often after a period of hours.

In recent years the precipitin test in liquid (usually normal saline) medium has been largely replaced by precipitin reactions in agar gel, Devised by Onchterlony (1950), this method is known as the plate test. In this method agar gel is prepared in a petri plate with a number of wells in it. Usually one well is bored in the center and six or seven in the peripheral regions. The known antigen is placed in the central well. The unknown and a suitable control are placed in the peripheral ones. The preparation is kept moist and within a few hours the protein solutions diffuse through gel. A clear line of precipitate is produced where antigen and antibody meet. The precipitate can be made clearly visible by staining with a dye.

Of the other tests, the complement fixation test (CFT) and anaphylaxis find extensive application in animal virology, particularly medical virology. The role of CFT in elucidation of the problem of possible viral oncogenesis cannot be over emphasized.

Serological data have also been utilized for finding out relatedness or otherwise amongst viruses. As early as in 1937 Bawden and Piric reported that the viruses then known as Cucumber viruses 3 and 4 were in fact related to tobacco mosaic virus. On the other hand, they reported that tobacco necrosis virus was in fact comprised of a number of serologically distinct virus types. Since then serological relationships between otherwise considered distinct virus types have been reported from time to time. For instance, the serological relationship between pea mosaic and bean yellow mosaic viruses.

Taxonomy of Viruses

Viruses are separated into families on the basis of the type ad form of the nucleic acid genome and the size, shape, substructure, and mode of replication of the virus particle. Within each family, classifications of genera and species are based on antigenicity in addition to other properties.

Significant developments in classification and nomenclature of viruses are documented in the reports of the International Committee on Taxonomy of Viruses, formerly the International Committee on Nomenclature of Viruses. These reports, published in 1971, 1976, 1979 and 1982, have dealt with viruses of human, lower animals, insects, plants, and bacteria and have included summaries of the properties of those groups of viruses as related to their taxonomic placement. It seems probable that most of the major groups of viruses have been recognized, particularly with regard to those infecting humans and the vertebrate animals of direct importance to humans. Many of them now have been officially placed in families, genera, and species; within some families, subfamilies or subgenera or both also have been established. Although the focus of this chapter is on these families, progress also has been made with respect to the taxonomy of the viruses of other host groups.

Properties of the major families of RNA containing viruses of humans and other vertebrate animals, vertebrate viruses with a DNA genome are similarly treated. These families are listed along with genera and the individual members that may be of special concern for the viral diagnostic laboratory. Where subfamilies have been designated, these also are included if they represent agents having direct or indirect bearing upon human diseases. Since most of these virus agents are dealt with more fully in the other chapters of this volume, the brief text which follows is confined to explanation and commentary on some examples from the tables, together with some notes about recent developments.

Picornaviridae

Of particular importance is the recent classification of hepatitis A virus as enterovirus 71, within the family Picornaviridae. This virus has been shown to have the physicochemical properties of a member of the genus *Enterovirus*. These properties include a nonenveloped icosahedral (cubic) virion about 27 nm in diameter, a buoyant density in CsCl of ca. 1.33 to 1.34 g/cm^3, and four major polypeptides with molecular weights of about 33,000, 27,000, 23,000, and 6,000. The genome consists of a single piece of single stranded RNA of molecular weight ca. 2.5 $\times$ 10^6. Like the other enteroviruses, hepatitis A virus is stable to acid $_p$H and resistant to ether. In its resistance to thermal inactivation this serotype differs somewhat from other enteroviruses. In comparative studies, 50% of the particles in poliovirus type 2 preparations disintegrate during heating at pH 7 for 10 min at 43°C, whereas under the same conditions 61°C is required to produce disintegration of 50% of the hepatitis A virus (enteroviruses, is stabilized by $MgCl_2$ against thermal inactivation. In the presence of 1M $MgCl_2$ the temperatures required to produce 50% destruction of the particles are 61°C for poliovirus type 2 and 81°C for enterovirus 72. Stabilization of infectivity by $MgCl_2$ is a well-established characteristic of enteroviruses. Enterovirus 72 shows unusual thermal resistance with $MgCl_2$ (loss of only 1.3 logs of infectivity after heating at 65°C for 10 min), but in the presence of $MgCl_2$ it is even more resistant. There is no reduction after 20 min at 65°C, and only 1.5 log units of infectivity is lost after 10 min at 80°C.

Reoviridae

The RNA genome is single stranded except in the case of family Reoviridae, whose RNA is double stranded. The genus *Reovirus* differs somewhat from the other genera in its possession of an outer protein shell and the larger molecular weight of its genome (15×10^6 versus 12×10^6). There are three serotypes within the *Reovirus* genus that infect humans, monkeys, dogs, and cattle; in addition, at least five avian reoviruses are known. In the genus *Orbivirus*, important human pathogens include Colorado tick fever virus and the Kemerovo viruses; other important orbiviruses are the bluetongue viruses; other important orbiviruses are the bluetongue viruses of sheep and the viruses of African horse sickness. The most recently established genus is *Rotavirus*, which includes several viruses that infect humans as well as viruses of many other mammalian species, typified by simian virus SA-11 and Nebraska calf diarrhea virus. The human rotaviruses are increasingly being recognized as the cause of a large share of the serious episodes of nonbacterial infantile diarrhea. Rotavirus gastroenteritis is one of the most common childhood illnesses throughout the world and is a leading cause of infant deaths in developing countries. These viruses also infect adults, particularly those in close contact with infants and children, but infected adults usually experience no symptoms or may have only minor illness.

Caliciviridae

Other recent additions to the taxonomic roll of RNA-containing viruses are the Caliciviridae and the Bunyaviridae. The Caliciviridae included a number if viruses of pigs, cats, and sea lions and may include agents that infect humans. Calicivirus-like particles have been observed in human feces in association with gastroenteric disease; preliminary results have failed to show relationship to the feline calicivirus. The possible relationship of these agents to the virus of Norwalk gastroenteritis also remains to be resolved. The Norwalk virus, a widespread human agent causing acute epidemic gastroenteritis, has a virion protein structure similar to that of the caliciviruses; it also resembles caliciviruses in several other characteristics. Because these agents have not yet been successfully adapted to tissue culture, it has been difficult to study their properties.

Bunyaviridae

The Bunyaviridae form a family of more than 200 viruses, at least 145 of them belonging to the Bunyamwera supergroup of serologically interrelated arboviruses. With the taxonomic placement of this large group, the vast majority of the viruses of the classical arbovirus groupings, initially based on ecological properties and subdivided by serological interrelationships, have been assigned to families on the basis of biophysical and biochemical characteristics. Human illness caused by Hantaan virus has been recognized in the Far East as Korean hemorrhagic fever, and a variant is known in Scandinavian and eastern European countries as epidemic nephropathy. The illness has been variously named "hemorrhagic fever with renal syndrome" or muroid virus nephropathy." Recent studies show the causative agent to be a member of the Bunyaviridae. The virus has a labile membrane and a tripartite single-stranded RNA genome. The most common natural hosts are mice (in Korea) and voles (in Europe). There have been several instances of infection of staff members handling laboratory rats infected with the virus, both in the Far East and more recently in Europe. In Belgium there also have been sporadic cases with no apparent link to an outbreak or to each other.

Retroviridae

The family Retroviridae has been divided into subfamilies. The best-known retroviruses belong to the subfamily Oncovirinae, the RNA tumor virus group, which has been the focus of special interest because its members, long recognized as causing leukemia and sarcoma in animals, serve as valuable animal models of oncogenic viruses.

The Retroviridae are enveloped viruses whose genome contains a single-stranded RNA of the same polarity as viral mRNA. The virion contains a reverse transcriptase enzyme. Replication proceeds off an integrated "provirus" DNA copy in infected cells. Study of retroviruses, and oncoviruses in particular, has permitted the identification of cellular "oncogenes." Normal cells of several animal species contain integrated copies of the genes of the endogenous oncovirus. The oncovirus genes may not be expressed but can be activated by physical and chemical agents, by superinfection with other oncoviruses, and even by herpesviruses. Recently, members of the subfamily oncovirinae have been shown to cause human disease. These include the agents generally known as the human T-cell leukemia viruses and the agent known as the lymphadenopathy-associated virus. There are at least three types of human T-cell leukemia viruses, and type 3 apparently is identical with the lymphadenopathy-associated virus; this serotype seems to be the etiological agent of AIDS (acquired immune deficiency syndrome).

Parvoviridae

All of the virus families have their DNA genome in double-stranded form except the Parvoviridae, whose DNA is single stranded within the virion. The members of Parvoviridae are very small viruses. The molecular weight of the nucleic acid in the virion is relatively very low, 1.5×10^6 to 2.0×10^6 (as compared, for example, with 160×10^6 for DNA of poxviruses). Some members display resistance to high temperatures (60°C, 30 min).

The family Parvoviridae encompasses viruses of numerous species of vertebrates, including humans. Two members of the genus *Parvovirus*, the members of which are able to replicate independently, have been found to be associated with disease problems of human beings. Parvovirus B19 has been shown to cause a transient shutdown of erythrocyte production by killing the late erythroid progenitor cells. This shutdown presents particular problems for individuals already suffering from hemolytic anemias such as sickle cell anemia, causing aplastic crises. A virus names RA-1, which is associated with rheumatoid arthritis, is another newly identified member of the genus.

A host range mutant of feline panleukopenia parvovirus, known as canine parvovirus, induces acute enteritis with leukopenia in young and adult dogs as well as myocarditis in puppies. Infections with this virus have reached enzootic proportions around the world.

Several serotypes of adeno-associated viruses, belonging to the *Dependovirus* genus, are known to infect humans, but they have not been shown to be associated with any human disease. Members of this genus cannot multiply in the absence of a replicating adenovirus which serves as a "helper virus". The single stranded DNA is present within the virion as either plus or minus complementary strands in separate particles. Upon extraction, the plus and minus DNA strands unite to form a double-stranded helix.

Papovaviridae

Members of the family Papovaviridae have DNA in double stranded, circular form. The human representatives are the papilloma or wart viruses and the JC and BK viruses; these latter were isolated, respectively, from the brain tissue of patients with progressive multifocal leukoencephalopathy and from the urine of immunosuppressed renal transplant recipients. In addition, several isolated, which appear to be identical to simian virus 40 of monkeys, have also been isolated from patients with progressive multifocal leukoencephalopathy Papovaviruses produce latent and chronic infections in their natural hosts. Many of them produce tumors, particularly in experimentally infected rodents, thus serving as models for studying viral carcinogenesis. The viral DNA integrates into cellular chromosomes of transformed cells.

When simian virus 40 and adenoviruses replicate together, they may interact to form "hybrid" virus particles, in which a defective simian virus 40 genome is covalently linked to adenovirus DNA and is carried within an adenovirus capsid.

Hepadnaviridae

Ample evidence has accumulated for the formation of a new virus family. The name, Hepadnaviridae, reflects the DNA-containing genomes of its members and their replication within hepatocytes. These viruses have a circular DNA genome that is double stranded except for a region of variable length that is single stranded. In the presence of appropriate substrates DNA polymerase within the virion can complete the single-stranded region to its full length of 3,200 nucleotides. Hepatitis B virus of humans and three similar viruses found in woodchucks, Beechey ground squirrels, and Pekin ducks share many basic features. All members of the family share antigens as well as similar morphology and behavior in the infected host. Large amounts of excess viral coat protein are produced in the form of small 22-nm spherical and tubular particles (in the human virus, the antigen is known as hepatitis B surface antigen). The viruses replicate in the liver and are associated with acute and chronic hepatitis. More than 200 million persons are persistent carriers of the human virus and are at very high risk of developing liver cancer. The woodchuck hepatitis B virus also causes liver cancer in its natural host. Fragments of viral DNA may be found in the liver cancer cells of both species.

7

Morphology of Viruses

Viruses are perhaps the simplest form of life known—simple, at least, when compared structurally with prokaryotic and eukaryotic cells. They are able to exist and replicate despite their simple structure because, like several other groups of organisms previously discussed, they are obligate intracellular parasites. But here the comparison of the viruses with rickettsiae or chlamydiae ends. Viruses lack components essential for their own replication and must depend entirely upon the host cell to provide these missing factors.

For example, biologic syntheses require energy, and this energy is provided by ATP in the form of high-energy phosphate bonds, broken to release energy for biosynthetic reactions. Viruses require ATP for a number of very important processes, foremost of which is the replication of their own genetic information. Because viruses lack ATP generating systems, they rely solely on such a system provided by the host cell.

A virus carries its own genetic information in the form of either RNA or DNA. Virus replication requires expression of this information via the classical pathways of transcription and translation of viral messenger RNAs. However, viruses do not contain all, or for that matter, most of the structural elements required for protein synthesis, that is, ribosomes factors, and tRNAs. Therefore, viruses must utilize ways of subverting normal cellular mechanisms and processes for their own replication. The extent to which they rely on cellular factors for replication depends in large part upon the complexity of their own genetic structure. For example, the large and complex herpesviruses encode many functions that simply replace the host's functions, whereas retroviruses, which encode only three proteins, must rely almost totally on host functions for replication.

Another characteristic peculiar to viruses is that, unlike all other forms of life, viruses contain only one type of nucleic acid; in some cases it is RNA, and, in others, it is DNA, but never both. Minor exceptions to this statement can be seen in the very complex, DNA-containing poxviruses that apparently contain very small amounts of their own RNA and in the RNA-containing retroviruses, which may contain trace amounts of DNA. However, in both cases, these trace amounts of "contaminating" nucleic acid are probably not necessary to the virus. The type of nucleic acid contained in a virion greatly influences the nature of its replicative pathway and, hence, is a very useful consideration when classifying viruses.

Sizes of Viruses

Viruses vary considerably in size, but in general, they are well below the limit of visibility of the light microscope. In early literature, they were often referred to as filterable agents, because they would pass through filters that would normally retain bacteria. In fact, early errors concerning the etiology of a disease have occurred when a virus was considered to be the causative agent merely because the infectious agent was filterable.

Virus sizes range from 20 nm to 250 nm (nm = 10^{-9} meter, or one thousandth of a μm), although, as will be seen, such measurements can be misleading for helical shaped viruses such as the rhabdoviruses. Classically, three basic techniques are used to determine virus size: (1) filtration through graded collodian membranes for which the size of the pores in each membrane is known; (2) high speed centrifugation (greater than 100,000 times gravity), which provides data for the calculations of virus size based on the rate at which a particle of known shape will travel toward the bottom of centrifuge tube at a known centrifugal force; and (3) direct observation using an electron microscope, often including objects of known size in the preparation of comparison. Modern day instrumentation makes the latter method of choice today.

Chemical Composition and Structure of Viruses

Viruses viewed in their simplest form are nothing more than nucleic acid (either DNA or RNA) surrounded by a protein overcoat called a *capsid*. This simple characteristic becomes more complex because many viruses contain additional structural proteins and enzymes.

Also, some may have carbohydrates bound to the surface proteins (glycoproteins), and others may be enclosed in a membrane that is composed of virally coded proteins and host-cell lipids. Our discussion of virus structure begins with the characterization, function, and structure of a basic viral component, the viral capsid.

Viral Capsids

Nature has designed the viral capsid to performs several functions. The viral capsid protects the enclosed nucleic acid from both physical destruction and enzymatic hydrolysis by the extracellular milieu. It also provides binding sites that enable the virus to attach to specific receptor sites on the host cell. Electron microscopy has revealed that, with the exception of the relatively complex poxviruses, all animal viruses are either isometric or helical in shapes. Virtually all DNA-containing animal viruses are icosahedrons, an exception being brick-shaped poxviruses; RNA-containing animal viruses may exist as icosahedrons or as helical viruses.

Isometric viruses

Isometric viruses appear to be in the shape of an icosahedron. Each particle has 20 facets, each an equilateral triangle. These facets come together to form 12 vertices of 5 facets each. Schematic and electron microscopic views of such a virus. As depicted in this figure, the completed capsid is made up of repeating morphologic units called capsomers. These capsomers are visible by electron microscopy as small protein structures.

For some viruses, each capsomer may be composed of only a single repeating polypeptide, whereas in other viruses, each capsomer may consists of several different polypeptide molecules held together by noncovalent bonds to form the completed capsomer. Examination of figure will show that the capsomer at each of the 12 vertices is surrounded by only 5 other capsomers, whereas all other capsomers are adjacent to 6 other capsomers. The capsomers at the vertices are

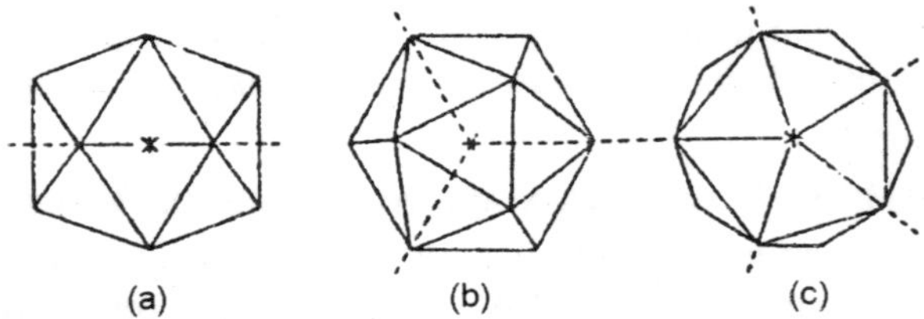

Fig. 7.1. Features of icosahedral symmetry. Depending upon the angle of observation, the icosahedron will show two, three, or five axes of symmetry.

called pentons, or pentomers, and the remaining capsomers are referred to as hexons, or hexomers. A special property of an icosahedral structure is its multiple axes of symmetry is possible to discern three different types of symmetry as one rotates the icosahedron, through the edges, the faces, and the vertices, resulting in what is referred to as 5:3:2 symmetry.

One means of classifying viruses is based on the number of capsomers present in the viral capsid. This number can be calculated for many isometric virions by using the formula $N=10\ (n-1)^2 + 2$, where n is the number of capsomers on one side of each equilateral triangle. Examples that fit this formula would include herpesvirus, with 5 capsomers making up the intact capsid, and adenovirus, with 6 capsomers per side and a total of 252 capsomers per intact virion.

Helical viruses

In the case of helical viruses, the viral nucleic acid is closely associated with the protein capsid, forming a coil-shaped nucleocapsid that becomes enclosed in a membrane as it buds from the host cell. The exact nature of the nucleic acid interaction with the capsid is not well understood, but the protein protects the RNA from enzymatic degradation by nucleases while still allowing the RNA to be transcribed from the intact nucleoprotein. Figure shows electron micrographs of rhabdovirus virions in which the striations of nucleic acid can be seen wound in a helical pattern in the interior of the particles. With the exception of the bullet-shaped rhabdoviruses, other helical animal viruses exist as spherical virions containing a helical-shaped nucleocapsid surrounded by a membrane.

Viral Nucleic Acid

Each family of viruses possesses a nucleic acid characteristic for that group, which, along with the properties of the virion capsid, is used for the classification of the virus. Viruses may possess double stranded DNA (ds-DNA), single-stranded DNA (ss DNA), ds RNA, or ss RNA. Furthermore, the amount of nucleic acid present may vary from enough to code for only three or four proteins (e.g., parvoviruses, picornaviruses) to that seen in the herpesviruses and poxviruses, which can code for hundreds of proteins (e.g., parvoviruses, picornaviruses) to that seen in the herpesviruses and poxviruses, which can code for hundreds of proteins.

Structure of viral DNA

Differences in viral DNA structure dictate that viruses involve different strategies for replicating their genetic information. As will be seen later, the elucidation of these mechanisms by the molecular biologist has yielded invaluable insights into the mechanisms of cellular DNA replication and has provided model systems for the analysis of DNA replication.

Supercoiling

The DNA of most animal viruses exist as a linear molecule, either ds DNA or ss DNA, or a covalently closed, supercoiled circle. Supercoiled DNA results from extra turns in the helical structure of ds DNA which are introduced by an enzyme called DNA gyrase. Supercoiled DNA can be converted into a simple circular structure by enzymatically breaking (nicking) one strand of the ds DNA. This relieves the supercoiled twist, permitting the relaxation of the molecule to form a "relaxed" circular molecule. Supercoiling occurs in all types of DNA, mitochondrial DNA, chloroplast DNA, and bacterial plasmid DNA as well as viral DNA. It undoubtedly provides an important organizational function for certain DNA molecules by permitting a higher-order structuring of the DNA.

Terminal replications

Molecular biologists have long appreciated the fact that bacteriophage DNAs frequently have unique sequence arrangements at their ends. Bacteriophage λ contains complementary termini which make it possible for the DNA to form circular structures during the process of DNA replication. The termini of the DNA of many viruses, such as adenoviruses and parvoviruses, contain inverted repeat sequences. The terminal sequence on each strand of DNA may be represented as 5'-ABC... C'B'A' (where ABC are complementary to A'B'C'). Such sequence organization can be visualized in the electron microscope if the ds DNA is heated and the separated strands are permitted to reanneal. In such a case, part of the DNA would base-pair to reform the original ds DNA, and part would base-pair its 3' and 5' ends so as to form a panhandle or lollipop structure similar to that described for a transposon. Herpesvirus DNA possesses a more complicated organization of repeated DNA sequences. Sequences complementary to the terminally repeated sequences are also found within the viral DNA sequence, providing an internal repetition.

The existence of terminal redundancy in viral DNA provides a unique structural feature important for the replication of the virus. Such terminal redundancy makes it possible for the same protein to recognize both ends of ds liner DNA. In addition, inverted terminal repetitions in ss DNA may provide a double-stranded region necessary for the initiation of replication of the ss DNA.

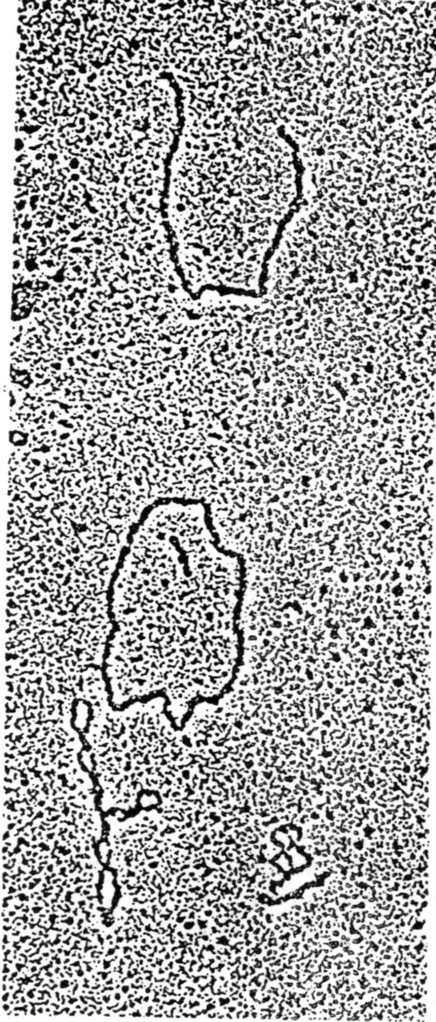

Fig. 7.2. As shown in this electron micrograph of SV40 DNA, isolated viral DNA may be obtained as linear, circular, or coiled and supercoiled molecules.

Structure of viral RNA

The RNA in animal viruses exists only as linear double-strand or single-strand molecules. Unlike DNA, however, the RNA within a virion may exist as a segmented genome. For example, reoviruses contain 10 different molecules of ds RNA; rotaviruses contain 11 segments of ds RNA, and influenza viruses possess 8 separate segments of ss RNA. Other RNA viruses containing a segmented genome include the bunyaviruses and arenaviruses. Retroviruses contain two identical single-strand genomes. Such a complex organization of genetic information requires unique mechanisms to ensure the proper segregation of genetic material and also provides the viruses with unique opportunities for the alteration of the genetic composition of the virus.

Plus and Minus strands of nucleic acid

Viruses containing single-stranded nucleic acid package only one of the two strands within the virion. A useful convention has been established to distinguish viruses containing ss RNA of different polarities. Those viruses containing ss RNA, which acts directly as messenger RNA (mRNA), have been designated as plus (+)-strand viruses, whereas those that must first replicate their RNA to form a complementary strand (which acts as the mRNA) are designated as minus (–)-strand viruses. As one might expect, the naked RNA of plus-strand viruses, such as poliovirus, can infect an animal cell, directing the synthesis of complete virus particles, because the entire message is carried on the mRNA, which is ready by the host-cell enzymes. Naked RNA from minus-strand viruses, however, is not infectious because it must first be replicated to form the complementary mRNA. Such replication is always catalyzed by an RNA polymerase carried within the virion.

DNA viruses can also exist as plus or minus ss DNA, but as DNA viruses must be converted to ds DNA in order to be transcribed by RNA polymerase to form mRNA. During transcription of ds DNA, generally only one strand is read, and one can consider that strand to be plus DNA and its complementary strand to be minus DNA. Classification of a virus as either (+) or (–) strand virus is extremely useful because it immediately reveals certain features of the strategy of virus replication.

Viral Envelopes

We have thus far described a virion as consisting of a protein capsid enclosing the viral nucleic acid and, in some cases, one or more vial enzymes. This combination of capsid and nucleic acid is referred to as the nucleocapsid, and, in the case of some animal virus groups, this constitutes the completed virion. A large number of animal viruses, however, contain a capsid surrounded by a lipid envelope acquired during the final stage of replication as they bud through special areas of the host cell membrane. These special areas of host cell

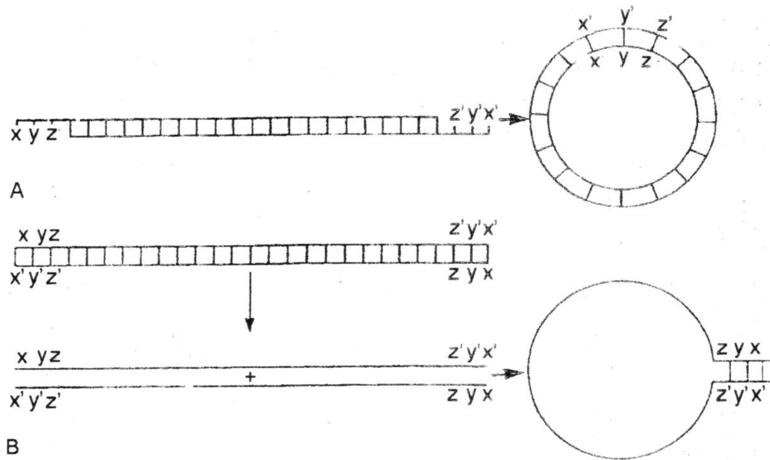

Fig. 7.3. Viral DNAs can have different terminal sequences. A–This shows the base-pairing between the ends of a DNA molecule with cohesive termini. B–This shows the possible base-pairing between the ends of a DNA molecule containing an inverted repeated sequence.

membrane are areas in which the host cell proteins have been supplanted by viral-coded polypeptides and virus-specified glycoproteins, although it is likely that the carbohydrate moiety of the glycoprotein usually occur as projections or spikes on the outer surface of the envelope. These protrusions have been given the name peplomers. Many of these glycoproteins have special functions, in particular, mediating the attachment of the virus to host cell receptors to initiate the entrance of the virion into the cell. Some glycoproteins will also attach to receptors on red blood cells, causing these cells to agglutinate. Others possess enzymatic activity and will cleave neuraminic acid from host cell glycoproteins. It is not surprising, therefore, that antibodies capable of neutralizing viral infectivity are frequently specific for the viral glycoprotein spikes.

In addition to protein and glycoprotein, the viral envelope contains 20% to 30% lipid, totally derived from the host cell membrane. The presence of this lipid makes it simple to determine whether or not a virion possesses an envelope. The infectiousness of enveloped viruses is inactivated by lipid solvents such as ether, whereas most viruses existing as naked nucleocapsids are resistant to such treatment. Thus, in many respects, the viral envelope is structurally analogous to the host cell membrane, differing only in that it contains virus-coded proteins and virus-specified glycoproteins.

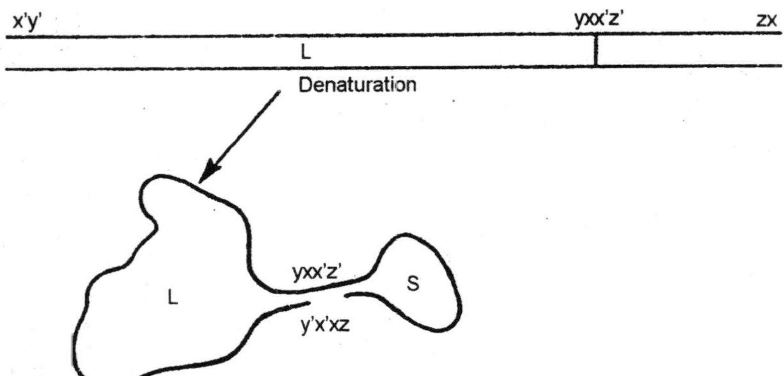

Fig. 7.4. Schematic representation of the herpesvirus genome.

Nonstructural Viral Proteins

A number of viruses carry within their capsids one or more enzymes that are liberated after the virus is uncoated in the host cell. By far the most common viral enzymes are viral-encoded polymerases. With the exception of the (+)-strand viruses, all RNA viruses must carry their own RNA polymerase. Without a viral polymerase, the viral RNA could not be transcribed and the virus could not replicate. The plus-strand RNA viruses encode their own RNA polymerase, which is synthesized by host cell translation of the viral mRNA.

Other enzymes are found in some larger DNA viruses; these are discussed when the replication of these viruses is described.

Classification of Animal Viruses

As our knowledge of viruses has increased, the properties used to classify them into families, groups, or genera have shifted from the symptomology or pathology of the infections they cause to the physical characteristics of the virion itself. To standardize the naming of new viruses, an International Committee on Nomenclature of Viruses (ICNV) was established at the Ninth International Congress for Microbiology in 1966. During the past decade and a half, this committee has set forth the following criteria for the classification of viruses: (1) the chemical nature of the nucleic acid (that is, DNA or RNA, single- or double-stranded, single or segmented genome, approximate molecular weight of nucleic acid, and whether the virion contains a plus or minus strand of nucleic acid); (2) the symmetry of the nucleocapsid (i.e., isometric or helical); (3) the presence or absence of an envelope; and (4) the number of capsomers for isometric virions, or the diameter of the nucleocapsids for helical viruses.

Based on these criteria, most of the viruses infecting mammalian hosts have been assigned to one of a number of virus families, and these in turn have been subdivided into one or more genera.

It should be emphasized to readers new to the field of medical virology that, although a generic-type classification of viruses is essential to provide a systematic handle for the animal viruses, some of the generic names are new and may be rarely used, especially by the medical virologist, who is more likely to refer to many of the viruses by the name of the disease for which they are responsible. This text briefly introduces the generic terminology, but our discussions for the most part use those virus names that are more commonly used by the medical virologist.

The following two sections, therefore, list the families of animal viruses along with a brief description of the ones that are discussed later in this text.

Families of DNA Viruses

Adenoviridae

Adenoviruses are large icosahedral viruses that cause various infections in humans. There are 41 known human adenoviruses, some of which will produce malignant tumors if injected into newborn hamsters. There is no evidence that these viruses will cause cancer in humans. They appear to persist in tissues for long periods of time and, in fact, were originally isolated from seemingly normal human adenoids, as suggested by their name.

Parvoviridae

Parvoviruses are small icosahedral viruses. This is the only group of animal viruses that contain single-stranded DNA. Subgenus A of the genus *Parvovirus* infects mice, rats, and swine.

Subgenus B contains defective viruses called adeno-associated viruses; these viruses are unable to replicate even within susceptible cells and can grow only if they coinfect a cell with a helper adenovirus. Thus, they are found in association with adenoviruses but are not known to cause disease in humans.

Papovaviridae

Papovaviruses are small, double-stranded DNA viruses, some of which have the ability to induce tumors. Human wart virus (*Papillomavirus*), Polyomavirus, and simian virus 40 (SV40) belong to this family. The papillomaviruses cause warts on the skin or on mucous membranes in humans, rabbits, cows, pigs, dogs, and other animals.

Structurally related viruses (BK, JC virus) have been isolated from humans. Polyomaviruses are known to produce a variety of cancers in susceptible animals such as mice and hamsters and are associated with certain malignant tumors in man.

Herpesviridae

Herpesviruses are large, enveloped icosahedral viruses that cause a rather wide variety of infections in humans and other animals. The most common human diseases caused by these viruses include fever blisters, chickenpox, and zoster (also known as shingles). Cytomegalovirus, a herpesvirus, causes serious congenital infections frequently manifested by mental retardation and defects of the central nervous system. Some herpesviruses are known to induce cancer in chickens, frogs, monkeys, and rabbits. Although there has been much speculation concerning the role of herpesvirus in human cancer, no indisputable proof is available at this time. A herpesvirus may be involved in a type of human cancer called *Burkitt's lymphoma*, and curiously, this same herpesvirus appear to be the etiologic agent for infectious mononucleosis. In recent years, herpesviruses have become a serious veneral disease.

Poxviridae

The poxviruses constitute a very large group of complex viruses that have been subdivided into six genera based on the animal hosts that they infect. The human poxviruses (*Orthopoxvirus*) include the etiologic agent of smallpox as well as vaccinia and related viruses.

Families of RNA Viruses

Picornaviridae

The family Picornaviridae contains two genera that cause human disease. The members of the genus *Enterovirus* are transmitted to humans primarily by way of fecal-oral route and, although diarrhea may be the major result of most enteroviral infections, many can invade the bloodstream and infect the central nervous system, causing aseptic meningitis and occasionally paralysis. Common human enteroviruses include the polioviruses, coxsackie viruses, echoviruses (enteric cytopathic human orphan viruses) and the etiologic agents of human hepatitis A (enterovirus 72). The second genus of Picornaviridae, *Rhinovirus*, causes a great deal of misery in humans: it encompasses the major etiologic agents of the common cold. There are more than 100 different serotypes of rhinoviruses, which undoubtedly accounts for the "commonness" of the common cold.

There are two additional genera of the Picornaviridae (*Cardiovirus* and *Aphthovirus*), which are animal pathogens.

Togaviridae

Togaviridae is the name given to a family of single-stranded RNA viruses possessing icosahedral symmetry. All are surrounded by a lipoprotein envelope, hence the prefix *toga*, meaning coat or mantle. Include in this family are many of the arboviruses (arthropod-borne viruses). There are more than 250 different arboviruses, all of which are transmitted from animal to animal (including humans) by a blood-sucking arthropod vector. Mosquitos and ticks are the usual vectors, and birds are the reservoir for many of these viruses. A few of the more common arboviral disease are eastern equine encephalitis (EEE), western equine encephalitis (WEE) Venezuelan equine encephalitis (VEE) (genus *alphavirus*). St. Louis encephalitis (SLE), Japanese B encephalitis (JE), yellow fever and dengue are also included in this group of viruses (genus *flavivirus*). On the basis of physico-chemical similarities, rubella virus, the causative agent of German measles, is also classified with the togaviruses (genus *rubivirus*), although it is not carried by arthropods.

Orthomyxoviridae

Orthomyxovirus, until recently, has been the generic name for the viruses that cause influenza, but this genus has now been renamed *Influenzavirus*. These agents are enveloped virions possessing helical symmetry. Projecting from their envelope are two different kinds of peplomers (spikes). One is called a *hemagglutinin* (HA) because it will react with a glycoprotein present on the surface of red blood cells, causing them to agglutinate. The other is an enzyme called *neuraminidase*, which can liberate neuraminic acid from the red blood cell glycoprotein, causing the spontaneous elution of the virus from the agglutinated red blood cells. Another characteristic of the influenzaviruses is that their single-stranded RNA exists within the virion in eight pieces.

Paramyxoviridae

The Paramyxoviruses include several animal pathogens such as Newcastle disease virus, virus 5, and five serotypes of parainfluenza virus that cause a mild respiratory disease in humans. These enveloped viruses possess a helical symmetry, but they can be rather pleomorphic (forming occasional filamentous virions); like the influenzaviruses, they carry a hemagglutinin and a neuraminidase on their surfaces.

Paramyxovirus RNA exists as a single, long minus strand. The major human pathogens in this family are measles virus and respiratory syncytial virus.

Rhabdoviridae

Rhabdoviruses are enveloped helical viruses that are characteristically bullet shaped. The two best-studied members of this family are rabies virus and vesicular stomatitis virus (a pathogen for cattle). There are, however, many other rhabdoviruses, including a number of fish pathogens. The rhabdoviruses also carry the minus strand of RNA and, hence, their own RNA polymerase.

Bunyaviridae

Bunyaviruses have been known for a number of years as the Bunyamwera supergroup. This group comprises a large number of arthropod-borne viruses, but, unlike the togaviruses, the Bunyamwera supergroup is made up of helical viruses containing single-stranded RNA that exists in three segments.

Arenaviridae

Arenaviruses were originally grouped together because of their morphologic similarities, but little is known concerning their detailed structure. A common characteristic shared by the arenaviruses is that they all appear to contain host-cell ribosomes within the viral capsid and to contain two segments of ss RNA. The best-studied arenavirus is lymphocytic choriomeningitis virus, which characteristically produces an asymptomatic infection in mice and occasionally causes meningitis in humans. Much more serious diseases produced by this group of viruses are the hemorrhagic fevers found in south America.

Coronaviridae

Coronaviruses are enveloped RNA viruses that are so named because of the petal-like projections surrounding their envelopes. They have been isolated from humans and other animals with upper respiratory disease.

Retroviridae

Retroviridae is the family name given to the RNA tumor viruses that were formally called the leukoviruses or oncornaviruses. These viruses produce a variety of neoplasms in birds and rodents. Retrovirus infections in humans have now been clearly documented and one such virus (HTLV-III) (AIDS virus) is major health problem. The symmetry of the retrovirus capsid is not known, but the viral RNA consists of two identical pieces of (+)-strands RNA. One unusual property of these viruses is that they must first synthesize DNA complementary to the virion RNA before they can replicate. The virions accordingly contain an enzyme, called an RNA-dependent DNA polymerase. It is because of this unusual method of replication that these viruses have been named retroviruses.

Reoviridae

Reoviridae is a family of viruses that contains a genome of double-stranded RNA. The family comprises six recognized genera whose ds RNA is segmented into 10 to 12 pieces within the virion. Human reoviruses possess a double icosahedral capsid. They are involved in mild respiratory infections in humans but are also found in the human intestinal tract. Because they are frequently not associated with disease, they are given the designation "Reo" (Respiratory enteric orphan).

Members of *Orbivirus*, a genus of Reoviridae, possess an icosahedral capsid made up of 32 quite large, doughnut-shaped capsomers. They are primarily animal pathogens, causing diseases such as blue tongue in cattle and a serious diarrhea is newborn calves, pigs, and mice. A recently recognized virus that causes a severe human diarrhea, particularly in infants, also belongs to the family Reoviridae. This virus has been given the name of *rotavirus*. It should now be apparent that these "simple forms of life" are really not at all so simple and that in order to grow viruses in the laboratory, it is necessary to provide a source of living, susceptible host cells in which they can replicate. These procedures have become commonplace during the past several decades.

8

Classification and Nomenclature of Viruses

Outside their host cells, viruses survive as virus particles, also known as virions. The virion is a gene delivery system; it contains the virus genome, and its functions are to protect the genome and to aid its entry into a host cell, where it can be replicated and packaged into new virions. The genome is packaged in a protein structure known as a *capsid*.

Many viruses also have a lipid component, generally present at the surface of the virion forming an envelope, which also contains proteins with roles in aiding entry into host cells. A few viruses form protective protein occlusion bodies around their virions. Before looking at these virus structures we shall consider characteristics of the nucleic acid and protein molecules that are the main components of virions.

Virus Genomes

A virion contains the genome of a virus in the form of one or more molecules of nucleic acid. For any one virus the genome is composed of either RNA or DNA. If a new virus is isolated, one way to determine whether it is an RNA virus or a DNA virus is to test its susceptibility to a ribonuclease and a deoxyribonuclease. The virus nucleic acid will be susceptible to degradation by only one of these enzymes.

Each nucleic acid molecule is either single-stranded (ss) or double-stranded (ds), giving four categories of virus genome: dsDNA, ssDNA, dsRNA and ssRNA. The dsDNA viruses encode their genes in the same kind of molecule as animals, plants, bacteria and other cellular organisms, while the other three types of genome are unique to viruses. It interesting to note that most fungal viruses have dsRNA genomes, most plant viruses have ssRNA genomes and most prokaryotic viruses have dsDNA genomes. The reasons for these distributions presumably concern diverse origins of the viruses in these very different host types.

A further categorization of a virus nucleic acid can be made on the basis of whether the molecule is linear, with free 5' and 3' ends,

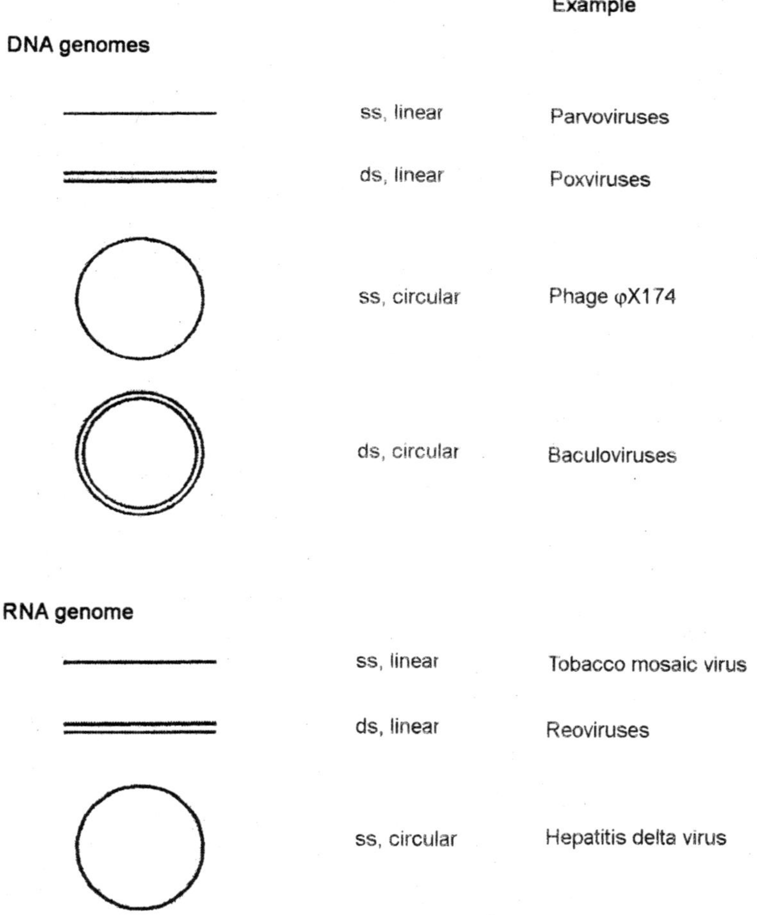

Fig. 8.1. Linear and circular viral genomes.

or circular, as a result of the strand(s) being covalently closed. Indeed throughout the book, molecules of DNA and RNA are colour coded. Dark blue and light blue depict (+) RNA and (–) RNA respectively.

It should be noted that some linear molecules may be in a circular conformation as a result of base pairing between complementary sequences at their ends. This applies, for example, to the DNA in hepadnavirus virions and to the RNA in influenza virions.

Genome Size

Virus genomes span a large range of sizes. Porcine circovirus (ssDNA) and hepatitis delta virus (ssRNA) each have a genome of about 1.7 kilobases (kb), while at the other end of the scale there are viruses with dsDNA genomes comprised of over 1000 kilobase pairs (kbp). The maximum size of a virus genome is subject to constraints, which vary with the genome category. As the constraints are less severe for dsDNA all of the large virus genomes are composed of dsDNA. The largest RNA genomes known are those of some coronaviruses, which are 33 kb of ssRNA. The largest virus genomes, such as that of the mimivirus, are larger than the smallest genomes of cellular organisms, such as some mycoplasmas.

Secondary and Tertiary Structure

As well as encoding the virus proteins (and in some cases untranslated RNAs) to be synthesized in the infected cell, the virus genome carries additional information, such as signals for the control of gene expression. Some of this information is contained within the nucleotide sequences, while for the single-stranded genomes some of it is contained within structures formed by intramolecular base pairing.

In ssDNA complementary sequences may base pair through G–C and A–T hydrogen bonding; in ssRNA weaker G–U bonds may form in addition to G–C and A–U base pairing. Intramolecular base pairing results in regions of secondary structure with stem-loops and bulges. In some ssRNAs intramolecular base pairing results in structures known as *pseudoknots*.

Regions of secondary structure in single-stranded nucleic acids are folded into tertiary structures with specific shapes, many of which are important in molecular interactions during virus replication. For example, the 5' end of poliovirus RNA, where there is an internal ribosome entry site to which cell proteins bind to initiate translation. Some pseudoknots have enzyme activity, while others play a role in ribosomal frameshifting.

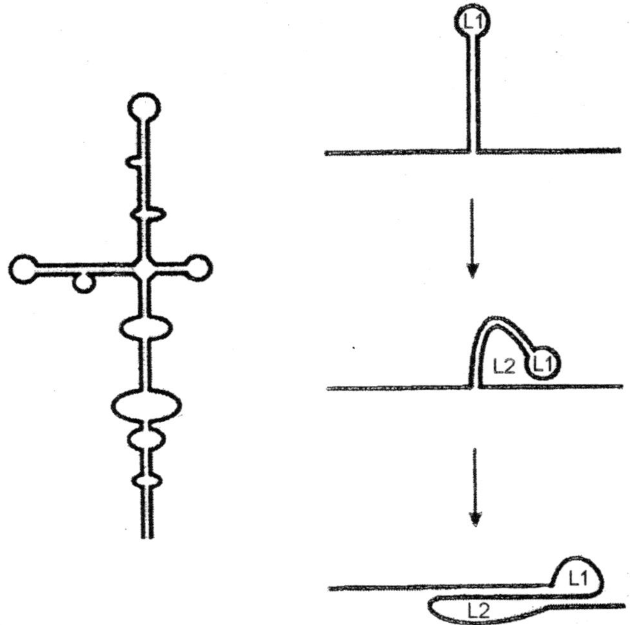

Fig. 8.2. Secondary structures resulting from intramolecular base-pairing in single-stranded nucleic acids. (a) Steam-loops and bulges in ssRNA and ssDNA. (b) Formation of a pseudoknot in ssRNA.

Modifications at the Ends of Virus Genomes

It is interesting to note that the genomes of some DNA viruses and many RNA viruses are modified at one or both ends. Some genomes have a covalently linked protein at the 5' end. In at least some viruses this is a vestige of a primer that was used for initiation of genome synthesis. Some genome RNAs have one or both of the modifications that occur in eukaryotic messenger RNAs (mRNAs): a methylated nucleotide cap at the 5' end and a sequence of adenosine residues (a polyadenylate tail; poly(A) tail) at the 3' end.

The genomes of many RNA viruses function as mRNAs after they have infected host cells. A cap and a poly(A) tail on a genome RNA may indicate that the molecule is ready to function as mRNA, but neither structure is essential for translation. All the ssRNAs function as mRNAs, but not all have a cap and a poly(A) tail. The genomes of some ssRNA plant viruses are base paired and folded near their 3' ends to form structures similar to transfer RNA. These structures contain sequences that promote the initiation of RNA synthesis.

Proteins Non-covalently Associated with Virus Genomes

Many nucleic acids packaged in virions have proteins bound to them non-covalently. These proteins have regions that are rich in the basic amino acids lysine, arginine and histidine, which are negatively charged and able to bind strongly to the positively charged nucleic acids. Papillomaviruses and polyomaviruses, which are DNA viruses, have cell histones bound to the virus genome. Most proteins associated with virus genomes, however, are virus coded, such as the HIV-1 nucleocapsid protein that coats the virus RNA; 29 per cent of its amino acid residues are basic. As well as their basic nature, nucleic-acid-binding proteins may have other characteristics, such as zinc fingers; the HIV-1 nucleocapsid protein has two zinc fingers. In some viruses, such as tobacco mosaic virus, the protein coating the genome constitutes the capsid of the virion.

Segmented Genomes

Most virus genomes consist of a single molecule of nucleic acid, but the genes of some viruses are encoded in two or more nucleic acid molecules. These segmented genomes are much more common amongst RNA viruses than DNA viruses. Examples of ssRNA viruses with segmented genomes are the influenza viruses, which package the segments in one virion, and brome mosaic virus, which packages the segments in separate virions. Most dsRNA viruses, such as members of the family *Reoviridae*, have segmented genomes.

The possession of a segmented genome provides a virus with the possibility of new gene combinations, and hence a potential for more rapid evolution. For those viruses with the segments packaged in separate virions, however, there may be a price to pay for this advantage. A new cell becomes infected only if all genome segments enter the cell, which means that at least one of each of the virion categories must infect.

Repeat Sequences

The genomes of many viruses contain sequences that are repeated. These sequences include promoters, enhancers, origins of replication and other elements that are involved in the control of events in virus replication. Many linear virus genomes have repeat sequences at the ends (termini), in which case the sequences are known as terminal repeats. If the repeats are in the same orientation they are known as *direct terminal repeats* (DTRs), whereas if they are in the opposite orientation they are known as *inverted terminal repeats* (ITRs). Strictly speaking, the sequences referred to as 'ITRs' in single-stranded nucleic acids are not repeats until the second strand is synthesized during replication. In the single-stranded molecules the 'ITRs' are, in fact, repeats of the complementary sequences.

Virus Proteins

The virion of tobacco mosaic virus contains only one protein species and the virions of parvoviruses contain two to four protein species. These are viruses with small genomes. As the size of the genome increases, so the number of protein species tends to increase; 39 protein species have been reported in the virion of herpes simplex virus 1, and over 100 in the virion of the algal virus *Paramecium bursaria*, *Chlorella* virus 1.

Proteins that are components of virions are known as structural proteins. They have to carry out a wide range of functions, including

1. protection of the virus genome
2. attachment of the virion to a host cell (for many viruses)
3. fusion of the virion envelope to a cell membrane (for enveloped viruses).

Virus proteins may have additional roles, some of which may be carried out by structural proteins, and some by non-structural proteins (proteins synthesized by the virus in an infected cell but they are not virion components). These additional roles include

1. enzymes, e.g. protease, reverse transcriptase
2. transcription factors
3. primers for nucleic acid replication
4. interference with the immune response of the host.

In a virion the virus genome is enclosed in a protein coat, known as a capsid. For some viruses the genome and the capsid constitute the virion, while for other viruses there are additional components. There may be an envelope at the surface of the virion, in which case there may be protein between the envelope and the capsid, or there may be an internal lipid membrane. A few viruses produce protein occlusion bodies in which virions become embedded. We shall consider each of these components in turn.

Capsids

Virus genomes removed from their cap sids are more susceptible to inactivation, so a major function of the capsid is undoubtedly the protection of the genome. A second major function of many capsids is to recognize and attach to a host cell in which the virus can be replicated. Although the capsid must be stable enough to survive in the extracellular environment, it must also have the ability to alter its conformation so that, at the appropriate time, it can release its genome into the host cell.

For many viruses the capsid and the genome that it encloses constitute the virion. For other viruses a lipid envelope, and sometimes another layer of protein, surrounds this structure, which is referred to as a *nucleocapsid*.

Capsids are constructed from many molecules of one or a few species of protein. The individual protein molecules are asymmetrical, but they are organized to form symmetrical structures. A symmetrical object, including a capsid, has the same appearance when it is rotated through one or more angles, or when it is seen as a mirror image. For the vast majority of viruses the capsid symmetry is either helical or icosahedral.

Capsids with Helical Symmetry

The capsids of many ssRNA viruses have helical symmetry; the RNA is coiled in the form of a helix and many copies of the same protein species are arranged around the coil. This forms an elongated structure, which may be a rigid rod if strong bonds are present between the protein molecules in successive turns of the helix, or a flexible rod if these bonds are weak. The length of the capsid is determined by the length of the nucleic acid. For many ssRNA viruses, such as measles and influenza viruses, the helical nucleic acid coated with protein forms a nucleocapsid, which is inside an envelope. The nucleocapsid may be coiled or folded to form a compact structure.

The virions of some plant viruses that have helical symmetry (e.g. tobacco mosaic virus) are hollow tubes; this allows the entry of negative stain, making the centre of the virion appear dark in electron micrographs. The rod-shaped tobacco rattle virus has a segmented genome with two RNAs of different sizes packaged in separate virions, resulting in two lengths of virion. The virions of a few DNA viruses, such as the filamentous phages, also have helical symmetry.

Capsids with Icosahedral Symmetry

Before proceeding further, a definition of the term '*icosahedron*' is required. An icosahedron is an object with

1. 20 faces, each an equilateral triangle;
2. 12 vertices, each formed where the vertices of five triangles meet;
3. 30 edges, at each of which the sides of two triangles meet.

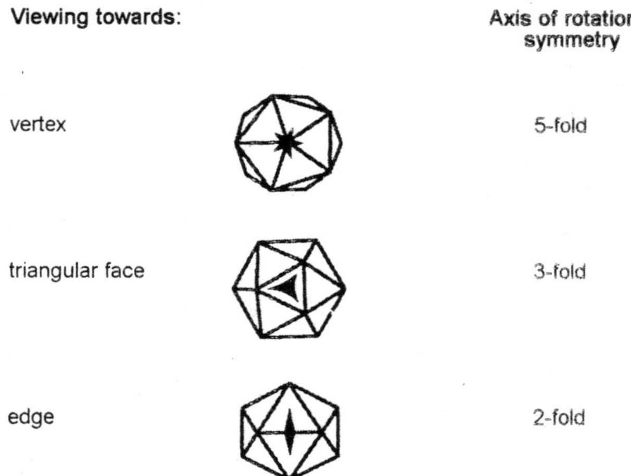

Fig. 8.3. The three axes of symmetry of an icosahedron.

An icosahedron has five-, three- and two-fold axes of rotational symmetry. Capsids with icosahedral symmetry consist of a shell built from protein molecules that appear to have been arranged on scaffolding in the form of an icosahedron. They have less contact with the virus genome than the capsid proteins of viruses with helical symmetry.

To construct an icosahedron from identical protein molecules the minimum number of molecules required is three per triangular face, giving a total of 60 for the icosahedron. The capsid of satellite tobacco mosaic virus is constructed in this way.

In capsids composed of more than 60 protein molecules it is impossible for all the molecules to be arranged completely symmetrically with equivalent bonds to all their neighbours. In 1962 Donald Caspar and Aaron Klug proposed a theory of quasi-equivalence, where the molecules do not interact equivalently with one another, but nearly equivalently. Hence, the capsid of a virus built from 180 identical

Fig. 8.4. Capsid constructed from sixty protein molecules—arrangement of protein molecules, with three per triangular face.

protein molecules, such as tomato bushy stunt virus, contains three types of bonding between the molecules.

The capsids of many icosahedral viruses are composed of more than one protein species. That of cowpea mosaic virus is composed of two proteins: one is present as 'pentamers' at the vertices of the icosahedron (12 × 5 = 60 copies) and the other is present as 'hexamers' on the faces. Each 'hexamer' is composed of three copies of a protein with two domains.

It was pointed out that there is a huge range in the sizes of virus genomes, with all the large genomes being dsDNA. There is also a huge range in the sizes of icosahedral capsids. The satellite tobacco mosaic virus capsid is about 17 nm in diameter, whereas the diameter of the *Paramecium bursaria Chlorella* virus 1 capsid is about ten times greater than this and the mimivirus capsid is about 300 nm in diameter.

Capsid shapes

It is clear that capsid surfaces vary in their topography; there may be canyons, hollows, ridges and/or spikes present. It is also clear that some capsids actually have the shape of an icosahedron, such as that of *Paramecium bursaria Chlorella* virus 1, which is 165 nm across when measured along the two- and three-fold axes and 190 nm across when measured along the five-fold axes. Capsids that have an icosahedral shape have an angular outline in electron micrographs.

An icosahedral shape is not an inevitable outcome of icosahedral symmetry; the football shape is constructed in the form of icosahedral symmetry, but the structure is spherical. Many small viruses that have capsids with icosahedral symmetry appear to be spherical, or almost spherical, and their virions are often described as isometric, such as those of densoviruses and foot and mouth disease virus.

Some capsids with icosahedral symmetry are elongated. The capsids of geminiviruses (plant viruses) are formed from two incomplete icosahedra. Another plant virus, alfalfa mosaic virus, has four sizes of virion; all are 19 nm diameter, but three are elongated as a result of insertions of a protein lattice between a half icosahedral structure at each end of the capsid.

Capsomeres

The capsids of some viruses, such as papillomaviruses, are clearly constructed from discrete structures. These structures are called capsomeres and each is built from several identical protein molecules. The capsids of papillomaviruses are constructed from 72 capsomeres, which are all identical, but the capsids of some viruses are constructed from two types of capsomere: pentons, which are found at the vertices of the icosahedron, and hexons, which make up the remainder of the capsid. In these viruses there are always 12 pentons (one at each vertex), but the number of hexons varies; for example, the capsids of herpesviruses and adenoviruses contain 150 and 240 hexons, respectively.

Structures at capsid vertices

Some icosahedral viruses have a structure such as a knob, projection or fibre at each of the 12 vertices of the capsid. For example, the virions of some phages have projections, while the adenovirus virion has a fibre, with a knob attached, at each of the 12 pentons. These structures at the capsid vertices are composed of distinct proteins that are involved in attachment of the virion to its host cell and in delivery of the virus genome into the cell.

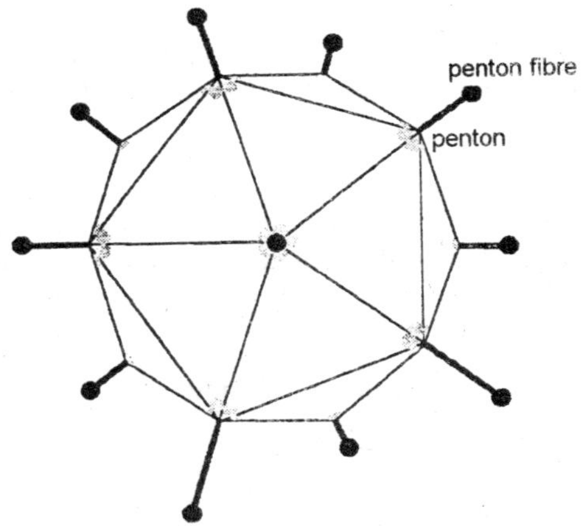

Fig. 8.5. Adenovirus virion.

Tailed bacteriophages

The majority of the known phages are constructed in the form of a tail attached to a head, which contains the virus genome. All of these phages have dsDNA genomes. The head has icosahedral symmetry and may be isometric as in phage lambda (λ), or elongated as in phage T4. The tail, which is attached to one of the vertices of the head via a connector, may be long as in phage λ, or short as in phage T7. Attached to the tail there may be specialized structures such as fibres and/or a baseplate.

Some of the tailed phages have been objects of intensive study and a lot of the detail of their structures has been uncovered. One

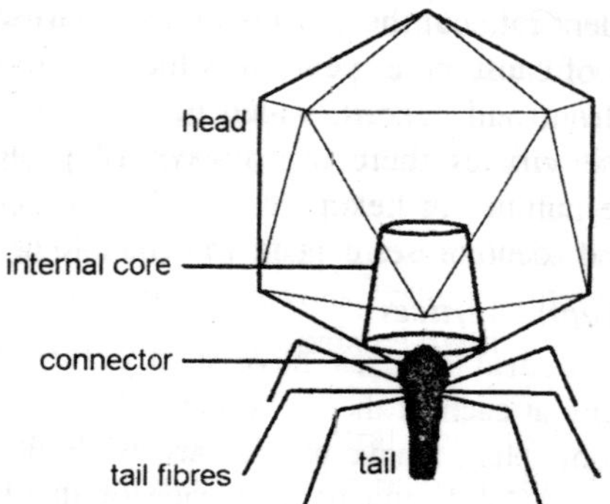

Fig. 8.6. Structure of phage T7.

such phage is T7. Inside the head of phage T7 is a cylindrical structure (the internal core) around which the DNA is wound. The connector has a wider region inserted into one of the vertices of the head and a narrower region to which the tail is attached. The tail is very short and tapers from the connector to the tip; attached to the tail are six tail fibres.

Conical and Rod-shaped Capsids

HIV-1 and baculoviruses have capsids that are conical and rod shaped, respectively. Inside each capsid is a copy of the virus genome coated in a highly basic protein. Both of these viruses have enveloped virions.

VIRION MEMBRANES

Many viruses have a lipid membrane component. In most of these viruses the membrane is at the virion surface and is associated with one or more species of virus protein. This lipid-protein structure is known as an envelope and it encloses the nucleocapsid (nucleic acid plus capsid). The virions of most enveloped viruses, such as herpesviruses, are spherical or roughly spherical, but other shapes exist. Some viruses have a membrane located not at the virion surface, but within the capsid.

Enveloped Virions

Many animal viruses are enveloped, including all those with helical symmetry, e.g. influenza viruses, and a significant number of those with icosahedral symmetry, e.g. herpesviruses. Enveloped virions are much less common amongst the viruses that infect plants (e.g. potato yellow dwarf virus) and they are extremely rare amongst the viruses that infect prokaryotes (e.g. *Pseudomonas* phage φ6).

Virion Shape	Examples
Enveloped Viruses	
Sphere	Influenza virus
Bullet	Rabies virus
Rod	Baculoviruses
Thread	Ebola virus
Viruses with an Internal Lipid Membrane	
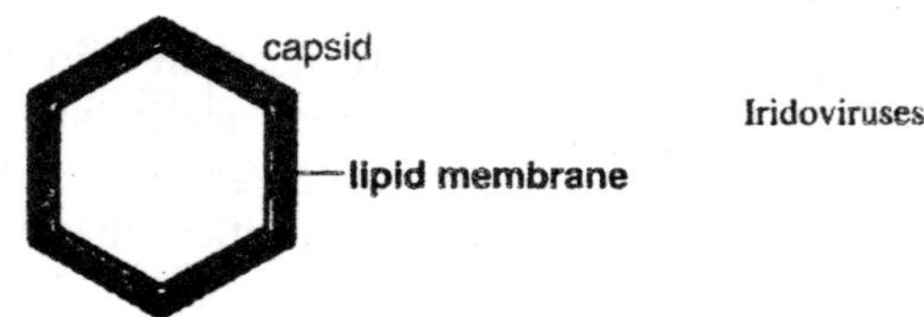	Iridoviruses

Fig. 8.7. Lipid-containing viruses.

Associated with the membranes of an enveloped virus are one or more species of protein. Most of these proteins are integral membrane proteins and most are O- and/or N-glycosylated. Many of the glycoproteins in virion envelopes are present as multimers. The envelope of influenza A virus, for example, contains two glycoprotein species: a haemagglutinin present as trimers and a neuraminidase present as tetramers. There is also a third protein species (M2) that is not glycosylated.

Some envelope proteins span the membrane only once, but some, such as those of hepatitis B virus, span the membrane several times. The polypeptide chain is highly hydrophobic at each membrane anchor.

Some surface glycoproteins of enveloped viruses perform the function of fusing the virion membrane to a cell membrane during the infection process. These fusion proteins have an additional hydrophobic region that plays a major role in the fusion process. Many enveloped viruses, including influenza viruses and retroviruses, have a layer of protein between the envelope and the nucleocapsid. This protein is often called a *matrix* protein. In some viruses, however, such as yellow fever virus, there is no such layer and the nucleocapsid interacts directly with the internal tails of the integral membrane protein molecules.

Virions with Internal Membranes

There are several viruses that have a lipid membrane within the virion rather than at the surface. These membranes have proteins associated with them and are present in the virions of, for example, *Paramecium bursaria Chlorella* virus 1, the iridoviruses and the tectiviruses (phages).

Membrane Lipids

Most virion membranes are derived from host cell membranes that undergo modification before incorporation into virions. For example, the HIV-1 envelope is derived from the plasma membrane of the host cell, but the virus envelope contains more cholesterol and sphingomyelin, and less phosphatidylcholine and phosphatidylinositol.

Some viruses are able to replicate in more than one kind of host cell; for example, alphaviruses replicate in both mammalian cells and insect cells. When progeny virions are released from cells the lipid composition of the envelope may reflect that of the cell. Virions of Semliki Forest virus produced in hamster kidney cells contain about five times more cholesterol than virions produced in mosquito cells.

Occlusion Bodies

Some viruses provide added protection to the virions whilst outside their hosts by occluding them in protein crystals. These occlusion bodies, as they are known, are produced by many of the viruses that infect invertebrates, including most baculoviruses. There are two major types of occlusion body in which *baculoviruses* embed their rod-shaped virions; the *granuloviruses* form small granular occlusion bodies, generally with a single virion in each, while the nucleopolyhedroviruses form large occlusion bodies with many virions in each.

Other Virion Components

Virus RNA in DNA Viruses

Virus RNA is present in the virions of a number of DNA viruses. The hepadnaviruses and the caulimoviruses have short RNA sequences covalently attached to their DNAs. These RNAs have functioned as primers for DNA synthesis and remain attached to the genome in the mature virion. There is evidence that virus mRNAs are packaged in the virions of herpesviruses, but the roles of these molecules are not clear.

Cell Molecules

The incorporation of cell lipids into virions has already been discussed. Other cell molecules that become incorporated into virions include the following.

1. *Transfer RNA molecules*. These are present in the virions of retroviruses.
2. *Proteins*. When a virion is assembled some cell protein may be incorporated. There are reports of HIV-1 incorporating several cell proteins, including cyclophilin A in association with the capsid and human leukocyte antigens in the envelope.
3. *Polyamines*. Spermidine and other polyamines have been reported in a variety of viruses.

Cations, such as Na^+, K^+, Zn^{2+} and Mg^{2+} have also been reported as components of virions. One likely role for polyamines and cations is the neutralization of negative charges on the virus genome.

Classification of Viruses

Virologists are no different to other scientists in that they find it useful to classify the objects of their study into groups and sub-groups. In the early days, when little was known about viruses, they were loosely grouped on the basis of criteria such as the type of host, the type of disease caused by infection and whether the virus is transmitted by an arthropod vector. As more was learnt about the characteristics of virus particles some of these began to be used for the purposes of classification, for example

1. whether the nucleic acid is DNA or RNA
2. whether the nucleic acid is single stranded or double stranded
3. whether or not the genome is segmented
4. the size of the virion
5. whether the capsid has helical symmetry or icosahedral symmetry
6. whether the virion is naked or enveloped.

Various combinations of these criteria produced some useful virus groups, but there was no single approach to the naming of the groups, and names were derived in a variety of ways:

1. small, icosahedral, single-stranded DNA viruses of animals were called parvoviruses (Latin *parvus* = small);
2. nematode-transmitted polyhedral (icosahedral) viruses of plants were called nepoviruses;
3. phages T2, T4 and T6 were called T even phages.

Serological relationships between viruses were investigated, and distinct strains (serotypes) could be distinguished in serological tests using antisera against purified virions. Serotypes reflect differences in virus proteins and have been found for many types of virus, including rotaviruses and foot and mouth disease virus.

International Committee on Taxonomy of Viruses

By 1966 it was decided that some order had to be brought to the business of naming viruses and classifying them into groups, and the International Committee on Taxonomy of Viruses (ICTV) was formed. The committee now has many working groups and is advised by virologists around the world.

The ICTV lays down the rules for the nomenclature and classification of viruses, and it considers proposals for new taxonomic groups and virus names. Those that are approved are published in book form and on the web; these sources should be consulted for definitive information.

Modern Virus Classification and Nomenclature

For a long time virologists were reluctant to use the taxonomic groups such as family, subfamily, genus and species that have long been used to classify living organisms, but taxonomic groups of viruses have gradually been accepted and are now established. Some virus families have been grouped into orders, but higher taxonomic groupings, such as class and phylum, are not used. Only some virus families are divided into subfamilies.

Each order, family, subfamily and genus is defined by viral characteristics that are necessary for membership of that group, whereas members of a species have characteristics in common but no one characteristic is essential for membership of the species. Many species contain variants known as virus strains, *serotypes* (differences are detected by differences in antigens) or *genotypes* (differences are detected by differences in genome sequence).

Many of the early names of virus groups were used to form the names of families and genera, e.g. the picornaviruses became the family *Picornaviridae*. Each taxonomic group has its own suffix and the formal names are printed in italic with the first letter in upper case, e.g. the genus *Morbillivirus*. When common names are used, however, they are not in italic and the first letter is in lower case (unless it is the first word of a sentence), e.g. the morbilliviruses.

Classification Based on Genome Sequences

Now that technologies for sequencing virus genomes and for determining genome organization are readily available, the modern approach to virus classification is based on comparisons of genome sequences and organizations. The degree of similarity between virus genomes can be assessed using computer programs, and can be represented in diagrams known as phylogenetic trees because they show the likely phylogeny (evolutionary development) of the viruses. Phylogenetic trees may be of various types.

1. Rooted - the tree begins at a root which is assumed to be the ancestor of the viruses in the tree.
2. Unrooted - no assumption is made about the ancestor of the viruses in the tree.

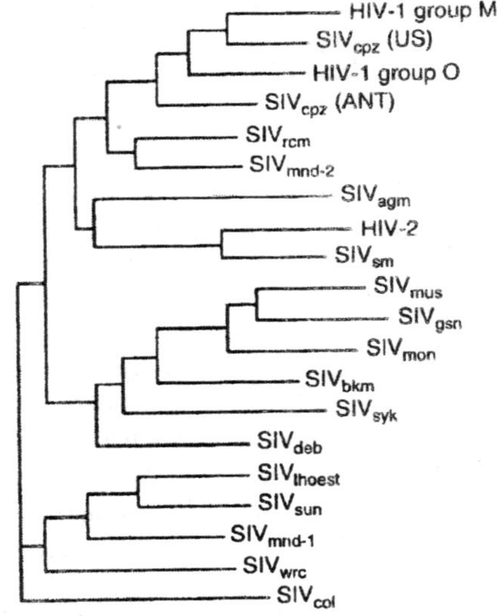

Fig. 8.8. Phylogenetic trees. Rooted tree of human immunodeficiency viruses (HIVs) and simian immunodeficiency viruses (SIVs) based on the RT-IN region of the pol gene.

The branches of a phylogenetic tree indicate how sequences are related. The branches may be scaled or unscaled; if they are scaled, their lengths represent genetic distances between sequences.

In many cases analysis of the sequence and organization of virus genomes has supported earlier classifications of viruses, e.g. the genera of the family *Reoviridae*. Another example is the rhabdoviruses, which were originally grouped together because of their bullet-shaped morphology, but it turns out that they are also related genetically.

Nomenclature of Viruses and Taxonomic Groups

The naming of individual viruses has been a rather hap-hazard business, with somewhat different approaches taken for viruses of different host types. Bacterial viruses were simply allotted codes, such as T1, T2 and φX174. Viruses of humans and other vertebrates were commonly named after the diseases that they cause, e.g. measles virus, smallpox virus, foot and mouth disease virus, though some were named after the city, town or river where the disease was first reported, e.g. Newcastle disease virus, Norwalk virus, Ebola virus. Some of these original names have been adopted as the formal names of the viruses.

Some of the place names where viruses were first found have become incorporated into the names of virus families and genera.

Table 8.1. Names of virus families and genera derived from place names

Place name	*Family/genus name*
Bunyanwera (Uganda)	Family *Bunyaviridae*
Ebola (river in Zaire)	Genus *Ebolavirus*
Hantaan (river in South Korea)	Genus *Hantavirus*
Hendra (Australia) and Nipah (Malaysia)	Genus *Hanipavirus*
Norwalk (United States)	Genus *Norovirus*

Many insect viruses were named after the insect, with an indication of the effect of infection on the host. A virus was isolated from *Tipula paludosa* larvae that were iridescent as a result of the large quantities of virions in their tissues. Another virus was isolated from *Autographa californica* larvae that had large polyhedral structures in the nuclei of infected cells. These viruses were named *Tipula* iridescent virus and *Auto grapha californica* nuclear polyhedrosis virus.

Most plant viruses were given names with two components: the host and signs of disease, e.g. potato yellow dwarf virus, tobacco rattle virus. Some of these names have been used as the bases for family and genus names.

Table 8.2. Names of families and genera of plant viruses based on the host and signs of disease

Host and disease signs	*Family/genus name*
Brome mosaic	Family *Bromoviridae*
Cauliflower mosaic	Family *Caulimoviridae*
Cowpea mosaic	Family *Comoviridae*
Tobacco mosaic	Genus *Tobamovirus*
Tobacco rattle	Genus *Tobravirus*
Tomato bushy stunt	Family *Tombusviridae*

As in other areas of biology, many names of virus taxonomic groups are based on Latin words, while some have Greek origins.

The student of virology thus gains some grounding in the classical languages! We can note that both Latin and Greek translations of 'thread' have been used to name the filoviruses and the closteroviruses, respectively. Both of these families have thread-shaped virions. Similarly, Latin and Greek translations of 'small' have been used to name the parvoviruses (animal viruses) and the microviruses (phages). The word for 'small' from a third language was used when devising a name for small RNA viruses; the Spanish '*pico*' was linked to 'RNA' to form 'picornaviruses'.

Baltimore Classification of Viruses

We have seen how viruses can be grouped into seven classes on the basis of the type of genome and the way in which the genome is transcribed and replicated. This approach to virus classification was first suggested by David Baltimore, after whom the scheme is named. An advantage of the Baltimore classification is its differentiation between plus-strand RNA viruses that do (class VI) and do not (class IV) carry out reverse transcription, and between dsDNA viruses that do (class VII) and do not (class I) carry out reverse transcription. In the chapters that follow we shall examine in depth a representative family of viruses from each of the seven Baltimore classes.

9

Method of Infection

A minimum proportion of the virions produced in infected hosts must be transmitted to new hosts in which more virions can be manufactured. If this does not happen the virus will die out. The only other possibility for the survival of virus genes is for them to be maintained in cells as nucleic acids, which are replicated and passed on to daughter cells when the cells divide.

Virus Transmission

Viruses of bacteria and other microbial hosts are released from infected cells into the environment of the host, where further susceptible cells are likely to be present. These viruses are dependent upon chance encounters with susceptible cells, to which they may bind if receptors on the surface of those cells come into contact with virus attachment sites. Viruses of multicellular animals and plants must also find new cells to infect. An infection may spread to adjacent cells, or to cells in a distant part of the host after transport in the blood of an animal or in the phloem of a plant, but ultimately a virus must find new hosts to infect if it is to survive.

Some virus infections modify the behaviour of their hosts in order to increase the probability of transmission. Mammals infected with rabies virus often become aggressive; this change in behaviour increases the likelihood of the host biting another individual and transmitting virus in the saliva. Some plant-feeding insect larvae infected with baculoviruses become more mobile in the late stages of infection, thus aiding virus dispersal. If the virus-infected insect migrates towards the top of a plant then lower leaves, where other larvae will feed, will become contaminated when the host dies and putrefies.

On their journeys between hosts viruses may have to survive adverse conditions in an environment such as air, water or soil. There are some viruses, however, that can be transmitted to new hosts without 'seeing the light of day', in other words without exposure to the outside environment. These are viruses that can be transmitted directly from host to host, for example during kissing or sexual intercourse, and viruses that are transmitted via vectors. Also included in this category are viruses that can be transmitted directly from a parent to members of the next generation. Transmission in these cases is said to be vertical; otherwise it is described as horizontal.

Viruses may be moved over long distances in a variety of ways. Rivers and winds can move viruses to new areas. The foot and mouth disease outbreak on the Isle of Wight, UK, in 1981 was initiated by virus that had spread via the air from Brittany, more than 250 km away. Viruses of birds, fishes, humans and other hosts are transported within their hosts to other parts of the planet as a result of migration, travel and animal export, for example

1. bird migration (avian influenza viruses)
2. human travel (SARS virus)
3. animal export (monkeypox virus).

Furthermore, viruses can be transported on inanimate materials, for example, foot and mouth disease virus on straw and farm vehicles.

Once virions have entered a multicellular organism they may have further to travel before suitable host cells are encountered. During this final stage of their journey the virions will encounter hazards in the form of host defence mechanisms, which must be survived if the virions are to remain infective when they reach their destination.

In theory, a single virion can initiate an infection, but in practice it is often found that a host must be inoculated with a minimum number of virions in order for that host to become infected. The reasons for this are probably many and varied, and may include some virions being defective and some being inactivated by the host's immune systems. This minimum amount of virus required for infection of a host is known as the minimum infective dose.

Transmission via Vectors: General Principles

Many viruses of plants and animals are transmitted between hosts by organisms that feed on them; these organisms act as vectors. There are several concepts concerning vector transmission that are common to viruses of both plants and animals so, before discussing these host categories separately, it is useful to consider these concepts.

Most vectors of viruses are arthropods (insects, mites, ticks); the arthropod-transmitted viruses of vertebrates are sometimes referred to as arboviruses (arthropod-borne viruses).

The principle of vector transmission is that the vector acquires a virus when it feeds on an infected host and subsequently transmits the virus to one or more new hosts. Some viruses are transmitted after virions have become attached to the mouthparts of their vectors during feeding. Transmission in this way may occur within seconds or minutes of the vector acquiring virus. Many vector-transmitted viruses, however, cross the gut wall of the vector and enter its circulatory system. The virus ultimately reaches the salivary glands and is secreted into the saliva, which may transport virus into new hosts when the vector feeds. This mode of transmission is said to be circulative, and transmission does not occur until hours or days after the vector has acquired the virus.

Some circulative viruses replicate in one or more tissues and organs of their vectors; thus there are viruses that can replicate in both invertebrates and plants, and viruses that can replicate in both invertebrates and vertebrates. It is fascinating to contemplate the existence of viruses with such broad host ranges. Many (but not all) of the invertebrates appear to suffer little or no harm when they are infected, so it has been hypothesized that many of the modern '*plant viruses*' and '*vertebrate viruses*' are descended from viruses of invertebrates that later extended their host ranges to plants or vertebrates.

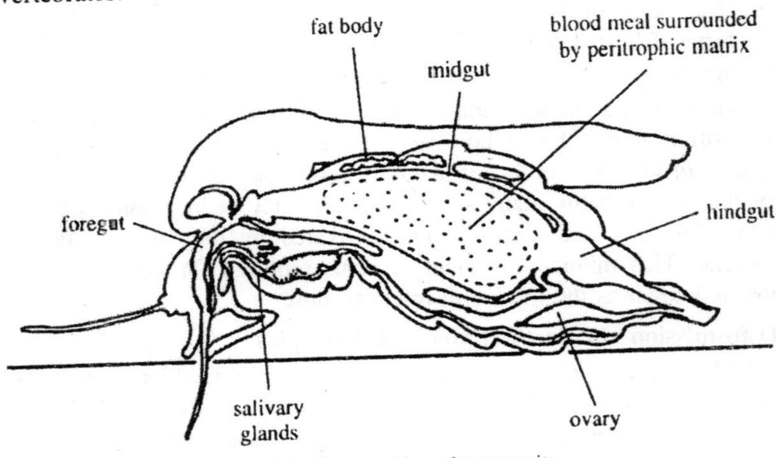

Fig. 9.1. Cross-section of a mosquito.

If the reproductive organs of the vector are infected there may be possibilities for vector-to-vector transmission. Some viruses are sexually transmitted (male to female and vice versa) and some are transmitted to the next generation within the egg; the latter is known as transovarial transmission. There is a high degree of specificity between most vectors and the viruses that they transmit. Generally specificity is greater between plant viruses and their vectors than between vertebrate viruses and their vectors.

Transmission of Plant Viruses

Plant cells are surrounded by thick cell walls that present significant barriers to virus entry; most plant viruses are carried across these barriers by vectors. A wide variety of organisms use plants as sources of nutrition and some of these organisms, especially invertebrates, act as virus vectors. Many of the vectors (e.g. aphids, nematodes) feed by piercing cell walls and ingesting the contents, while beetles feed by biting.

The most common vectors of plant viruses are aphids, which, as we have already pointed out, feed by ingesting the contents of cells. The reader may question how an invertebrate that feeds by removing the contents of plant cells can introduce virus into cells that subsequently support replication of the virus. The answer appears to be that these vectors probe a number of cells before selecting one on which to feed, so virus may be transmitted into cells that are probed, but not significantly damaged by the vector.

The nematodes that transmit viruses are soil- dwelling animals that pierce root cells and then ingest their contents. Virus transmission ceases after a nematode moults, indicating that virus does not move from its gut into its body. An interesting feature of virus- vector specificity concerns transmission of viruses with different shapes by different types of nematode. Tobraviruses have rod-shaped virions and are transmitted by nematodes in the family *Trichodoridae*, while nepoviruses have isometric virions and are transmitted by nematodes in the family *Longidoridae*.

The basis of some cases of plant virus-vector specificity lies in specific amino acid sequences in capsid proteins. In other cases important roles are played by virus-coded non-structural proteins (helper factors) that are synthesized in the infected plant cell. Virions may bind specifically to structures in the mouthparts of their vectors via specific sequences on the surfaces of capsids and/or helper factors.

Some plant-parasitic fungi can also act as virus vectors, for example *Spongospora subterranea*, which infects potato causing powdery scab disease, is a vector of potato mop-top virus. If a plant is infected with both a fungus and a virus, then virions may be taken into developing fungal spores. The virus may survive in a spore for months or years until it germinates on a new host, which then becomes infected, not only with the fungus but also with the virus. Some fungus-transmitted plant viruses have been classified in a genus, the name of which reflects their mode of transmission and the virion shape: the genus *Furovirus* (fungus-transmitted rod-shaped).

About 20 per cent of plant viruses can be transmitted vertically; in other words, seed can be infected, leading to infection of the next generation. Most seed-transmitted viruses are carried in the embryo, which may have acquired its infection from either an infected ovule or an infected pollen grain. Examples of viruses that can be transmitted via seed include the nepoviruses and the tobraviruses, two groups of nematode-transmitted viruses mentioned above. Many plant viruses can also be transmitted by artificial means; for example, grafting material from a virus-infected plant can introduce virus into a new host.

Transmission of Vertebrate Viruses

Non-vector Transmission of Vertebrate Viruses

Many viruses of vertebrates (including humans) infect their hosts via the mucous membranes of the upper respiratory tract. An infected individual may shed virus-containing droplets into the air during sneezing, speaking and coughing, and new hosts may become infected by inhaling this material. Virus present in genital secretions can be transmitted during sexual contact, while viruses infecting the intestinal tract are shed in faeces and may enter new hosts ingesting faeces-contaminated food or water.

Some viruses may be released from lesions, for example foot and mouth disease virus from lesions on the feet and in the mouth, papillomaviruses from warts, and herpes simplex virus from lesions on the lips. These viruses could be transmitted directly by contact between an infected host and a non-infected host, or they could be transmitted indirectly through contamination of the environment. Transmission of rabies virus, for example from dog to human, requires a bite from an infected mammal to introduce virus-containing saliva into the body.

In most of the cases discussed so far the host becomes infected when epithelial cells and/or lymphoid cells at a body surface become infected. Some infections (e.g. common cold, rotavirus) remain limited to these tissues, while others (e.g. measles, polio, HIV) cross the epithelial surface and spread to other organs and tissues.

Vector Transmission of Vertebrate Viruses

Vectors that transmit vertebrate viruses are mainly blood-feeding arthropods that acquire their viruses when they take blood meals from infected animals. Obviously for this to occur virus must be present in the blood of infected animals, a situation known as *viraemia*.

Some arthropod parasites that acquire virus infections from the vertebrates on which they feed remain infected for life. This is the case with ticks that become infected with tick-borne encephalitis virus. Transovarial transmission has been demonstrated for a number of viruses, including yellow fever virus in mosquitoes.

Viruses in blood or blood products can be transmitted during medical procedures by inanimate '*vectors*' such as blood transfusion equipment and syringes and needles. Viruses can also be transmitted when needles for drug injection are shared, when needlestick injuries occur and when haemophiliacs are injected with blood factors containing viruses.

Virus Survival in the Vertebrate Host

Once a virus has reached a potential host it must evade several defence systems if it is to initiate an infection. A virus in the lower respiratory tract must avoid removal by the mucociliary escalator, while a virus in the gastro-intestinal tract must avoid inactivation by extremes of pH. The mucosal surfaces of the respiratory, intestinal and urinogenital tracts present significant barriers to infection. The surfaces are bathed in fluids that contain antiviral substances and the fluids are viscous, limiting virus access to the cells.

The antiviral substances include complement proteins, which can be triggered by the presence of certain viruses to undergo modifications that help to protect the host from infection. Complexes of modified complement proteins can damage enveloped virions by insertion into their membranes. Some modified complement proteins can coat virions, which then attach to phagocytes via receptors on the cell surfaces.

Viruses may be phagocytosed by a range of cell types including macrophages, which are large cells. Viruses that are introduced into the blood by a vector may be phagocytosed by neutrophils, which constitute the majority of the white blood cells. These aspects of host immunity are not virus specific; they are components of the innate immune system. The virus may also have to contend with antibodies and other components of the adaptive immune system if the host has previously been infected with, or vaccinated against, the virus.

Transmission of Invertebrate Viruses

We have discussed the replication of 'plant' and 'vertebrate' viruses in invertebrate vectors, and it could be argued that these viruses are as much 'invertebrate' viruses as they are viruses of these other host types. In addition to these viruses, invertebrates are hosts to many other viruses that replicate only in invertebrates; most of the known invertebrate viruses have insect hosts.

Many of these viruses have evolved occlusion bodies, which are large protein structures in which virions become embedded in the infected cell. Viruses that produce occlusion bodies include cypoviruses and baculoviruses. Occlusion bodies may be expelled in the faeces of a virus-infected insect or they may remain in the host until it is ingested by a predator or it dies as a result of the virus infection. They are robust structures, able to survive putrefaction of the host and providing protection for virions in the outside environment. New insects become infected when occlusion bodies become ingested along with food, which is often plant material. Enzymes and high pH in the gut break down occlusion bodies, releasing the virions.

Vertical transmission and vector transmission of a number of invertebrate viruses have been reported. Vertical transmission may occur either within eggs (transovarially) or on the surface of eggs. Vectors involved in virus transmission between insect hosts include parasitic wasps.

Permissive Cells

In order to replicate itself, a virus must gain access to a permissive cell, meaning a cell that will permit the replication of the virus. For viruses that bind to the host cell surface as the first step in infection the cell must have appropriate receptors that the virus can bind to. Furthermore, all the requirements of the virus must either be present in the cell or they must be inducible; these requirements include cell proteins, such as transcription factors and enzymes.

Some viruses are restricted to a narrow range of permissive cell types, for example hepatitis B viruses are restricted almost exclusively to hepatocytes. Many other viruses are much less specific and, as we have seen in this chapter, some can replicate in both animal and plant hosts.

Experimentally, the range of permissive cell types may be much wider than the cell types found infected in nature. Flock house virus is an insect virus that was isolated from the New Zealand grass grub *Costelytra zealandica*. As well as replicating in other insect species after artificial inoculation, the virus can also replicate in a range of plant species. Some viruses of eukaryotes need the host cell to be in a particular phase of the cell cycle. For example, retroviruses require access to the cell nucleus and most can enter the nucleus only when the nuclear envelope has broken down in the M (mitosis) phase. Parvoviruses are DNA viruses that use the DNA replicating enzymes of the host to replicate their own DNA, so they need a cell in which those enzymes are present, i.e. a cell in the S phase.

A further requirement of a permissive cell is that it must lack defences against the virus, or the virus must have the ability to overcome the defences of the cell. Cells have evolved many mechanisms to defend themselves against infection with viruses and cellular microbes. Prokaryotic cells produce restriction endonucleases to degrade foreign DNA, including that of phages. (The cells modify their own DNA so that it is unaffected by the restriction enzymes.) The cells of vertebrate animals respond to the presence of dsRNA, which is produced during infection with many viruses, by synthesizing cytokines, including interferons, interleukins and tumour necrosis factor. These proteins can trigger a range of anti-viral defences. In order to successfully infect a cell a virus must deliver its genome into this cell, the genome must survive intact and all attempts by the host to terminate virus replication must be thwarted.

Attachment and Entry of Viruses into Cells

Overview of Virus Replication

The aim of a virus is to replicate itself, and in order to achieve this aim it needs to enter a host cell, make copies of itself and get the new copies out of the cell. In general the process of virus replication can be broken down into seven steps:

1. Attachment of a virion to a cell
2. Entry into the cell Transcription of virus genes into messenger RNA molecules (mRNAs)
3. Translation of virus mRNAs into virus proteins
4. Genome replication
5. Assembly of proteins and genomes into virions
6. Exit of the virions from the cell.

The first letter from each step gives the acronym AETTGAE, which may provide a memory aid. When trying to understand the modes of replication of different types of virus these steps provide a useful template. It is, however, a general template, because not all of the seven steps are relevant to all viruses, the steps do not always occur in the same order and some viruses have an additional step! In the later stages of replication several steps occur concurrently. For many viruses, transcription, translation, genome replication, virion assembly and exit can all be in progress at the same time. In this section we discuss the first two steps of this generalized replication cycle: attachment and entry into the host cell. We concentrate mainly on the mechanisms used by animal viruses to gain entry into host cells, but bacteriophages are also considered. The following three chapters deal with the remaining five steps of the generalized replication cycle.

The first step, attachment of a virion to a cell, applies to viruses that infect animal and bacterial hosts. Before these viruses can cross the bounding membrane or wall of the host cell they must first bind to specific molecules on the cell surface. Most plant viruses, on the other hand, are delivered directly into a cell by a vector.

Animal Viruses

Cell Receptors and Co-receptors

A virion attaches via one or more of its surface proteins to specific molecules on the surface of a host cell. These cellular molecules are known as receptors and the recognition of a receptor by a virion is highly specific, like a key fitting in its lock. It has been found that some viruses need to bind to a second type of cell surface molecule (a co-receptor) in order to infect a cell. In at least some cases binding to a receptor causes a conformational change in the virus protein that enables it to bind to the co-receptor.

Receptors and co-receptors are cell surface molecules, usually glycoproteins, with a wide range of functions that include

1. acting as receptors for chemokines and growth factors;
2. mediating cell-to-cell contact and adhesion.

A number of receptors are glycoproteins that are folded into domains similar to those found in immunoglobulin molecules. Many of the cell surface molecules used by viruses as receptors are in regions of the plasma membrane that are coated on the inner surface with one of the proteins clathrin or caveolin.

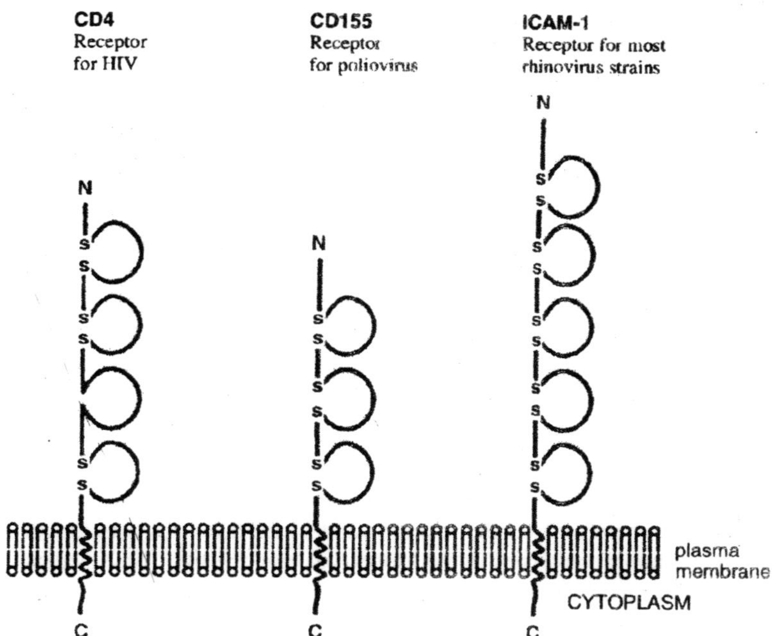

Fig. 9.2. Cell receptors with immunoglobulin-like domains.

Evidence that a cell surface molecule is a virus receptor

A search for the receptor for a particular virus might begin with the production of a panel of monoclonal antibodies against cell surface proteins. If one of the antibodies blocks virus binding and infectivity then this is strong evidence that the corresponding antigen is the receptor. Further evidence that a molecule is a virus receptor might come from experiments that show one or more of the following.

1. Soluble derivatives of the molecule block virus binding/infectivity.
2. The normal ligand for the molecule blocks virus binding/infectivity.
3. Introduction of the gene encoding the molecule into virus-resistant cells, and expression of that gene, makes those cells susceptible to infection.

Virus Attachment Sites

Each virion has multiple sites that can bind to receptors, and each site is made up of regions of one or more protein molecules.

The virus attachment sites of naked viruses are on the capsid surface, sometimes within depressions (e.g. poliovirus) and sometimes on ridges (e.g. foot and mouth disease virus). Poliovirus and foot and mouth disease virus are picornaviruses, and it has been demonstrated for picornaviruses in general that binding to receptors induces major structural changes in the virion.

The virus attachment sites of some naked viruses are on specialized structures, such as the fibres and knobs of adenoviruses and the spikes of rotaviruses, while the virus attachment sites of enveloped viruses are on the surface glycoproteins.

Some virion surface proteins that bear the virus attachment sites are able to bind strongly to red blood cells of various species and cause them to clump, a phenomenon known as haemagglutination. The proteins responsible for haemagglutination are called haemagglutinins. Examples of viruses that can haemagglutinate are influenza viruses and measles virus.

Attachment of Virions to Receptors

The forces that bind a virus attachment site to a receptor include hydrogen bonds, ionic attractions and van der Waals forces. Sugar moieties, when present on the receptor and/or on the virus attachment site, are commonly involved in the forces that bind the two. No covalent bonds are formed between virions and receptors.

Initially, a virion is weakly bound to a cell at only one or a few receptors. At this stage the attachment is reversible and the virion may detach, but if it remains attached there are opportunities for more virus attachment sites to bind to more receptors, and if sufficient bind then the attachment to the cell becomes irreversible.

Entry of Animal Viruses into Cells

After binding to receptors animal viruses must cross the plasma membrane to gain entry to the host cell. They may do this either at the cell surface or they may cross the membrane of an endosome, which is a vesicle formed by part of the plasma membrane pinching off into the cytoplasm. This process (endocytosis) is used by cells for a variety of functions, including nutrient uptake and defence against pathogens. There are a number of endocytic mechanisms, including clathrinmediated endocytosis and caveolin-mediated endocytosis; most animal viruses hi-jack one or more of these mechanisms in order to gain access to their host cells.

If a virion binds to a region of the plasma membrane coated with clathrin or caveolin these protein molecules force the membrane to bend around the virion. Many viruses, such as adenoviruses, are endocytosed at clathrin-coated regions of the plasma membrane. The virions end up in clathrin-coated endosomes, from which the clathrin is soon lost. Some viruses, such as simian virus 40, are endocytosed at caveolin-coated regions of the plasma membrane and the virions end up in caveolin-coated endosomes. Other viruses are taken up by mechanisms that are independent of clathrin and caveolin.

An endosome may fuse with other vesicles such as lysosomes, which have a pH of 4.8–5.0, thus lowering the pH within the vesicle. The pH may be further lowered by a process that pumps hydrogen ions across the membrane; the process involves the hydrolysis of ATP to ADP. This acidification of the environment of the virion is important for those enveloped viruses that need to carry out acid-triggered fusion of the envelope with the vesicle membrane.

Entry of naked viruses

It is possible that some naked viruses deliver their genomes into their host cells through a pore formed in the plasma membrane, but for most naked viruses irreversible attachment of the virion to the cell surface leads to endocytosis. The plasma membrane 'flows' around the virion, more receptors bind, and eventually the virion is completely enclosed in membrane, which pinches off as an endosome. The endosome contents, however, are part of the external environment and the virus is not yet in the cytoplasm. The mechanisms by which virions, or their genomes, are released from endosomes are not fully understood.

Entry of enveloped viruses

Reversible attachment of an enveloped virion may lead to irreversible attachment, as for naked viruses. There are then two processes whereby infection of the cell may occur: either fusion of the virion envelope with the plasma membrane, or endocytosis followed by fusion of the virion envelope with the endosome membrane.

Both processes involve the fusion of the virion envelope with a cell membrane, either the plasma membrane or a vesicle membrane. Lipid bilayers do not fuse spontaneously and each enveloped virus has a specialized glycoprotein responsible for membrane fusion.

Many fusion proteins are synthesized as part of a larger protein, which is cleaved after synthesis. The cleavage products remain linked, either through non-covalent bonds (e.g. retroviruses) or through a disulphide bond (e.g. influenza viruses), and are found in the virion envelope as dimers or trimers, depending on the protein. Each monomer has at least two hydrophobic sequences: a transmembrane sequence and a fusion sequence that is responsible for membrane fusion by inserting into a cell membrane.

The fusion sequence normally lies hidden and must become exposed in order for fusion to take place. Conformational changes to the fusion protein that expose the fusion sequence may come about either as a result of the virus binding to a receptor or they may be induced by the low pH within an endosome.

If the fusion sequence becomes exposed while the virion is at the cell surface, then infection could occur either by fusion with the plasma membrane or by endocytosis. If, however, exposure to low pH is required to expose the fusion sequence, then endocytosis is the only option. There are therefore two categories of membrane fusion:

1. pH-independent fusion, e.g. herpesviruses, HIV
2. acid-triggered fusion, e.g. influenza viruses, rhabdoviruses.

Once the fusion sequence is exposed it can insert into the target membrane (plasma membrane or endosome membrane). A further conformational change in the fusion protein pulls the two membranes together and then mediates their fusion. This process involves a release of energy from the fusion protein, which irreversibly changes shape.

Intracellular Transport

Once in the cell the virus, or at least its genome, may have to be delivered to a particular location, such as the nucleus. For some viruses the destination is reached using one of the transport systems of the cell, such as the microtubules.

Most RNA viruses of eukaryotes replicate in the cytoplasm; the majority encode all the enzymes for replication of their genomes and they have no requirement for the enzymes of the nucleus. The influenza viruses, however, are exceptions as they require the cell splicing machinery, so their genomes must be delivered into the nucleus.

Retroviruses too are RNA viruses that replicate their genomes in the nucleus. They copy their genomes to DNA in the cytoplasm, then most retroviruses must wait in the cytoplasm until mitosis begins. During mitosis the nuclear envelope is temporarily broken down and the virus DNA (with associated proteins) is able to enter the nuclear compartment. These viruses therefore can replicate only in cells that are dividing. The DNA (with associated proteins) of a group of retroviruses, however, can be transported into an intact nucleus. This group (the lentiviruses, which includes HIV) can therefore replicate in non-dividing cells.

Some DNA viruses, such as iridoviruses and poxviruses, replicate in the cytoplasm of eukaryotic cells, but most DNA viruses replicate in the nucleus. For these viruses (and the influenza viruses and the lentiviruses) the virus genome must be transported to the nuclear envelope and then across it. The structural proteins of some of these viruses have sequences that allow them to attach to microtubules.

Microtubules are components of the cytoskeleton, providing support for various components of the cell and acting as tracks for the transport of materials, such as certain organelles, to particular sites in the cell. Microtubules are hollow cylinders, 25 nm in diameter, and are composed of the proteins α- and β-tubulin. The ends of each microtubule are designated as plus and minus. In most animal cells the plus ends are located near the plasma membrane, while the minus ends are attached to a structure called the centrosome (also referred to as the microtubule organizing centre) near the nucleus.

Proteins, known as motor proteins, move themselves and any cargo along the microtubules. A number of viruses (including herpesviruses, adenoviruses, parvoviruses and retroviruses) exploit this transport system to take their nucleocapsid, or a structure derived from it, from the periphery of the cytoplasm to a location close to the nucleus. The virus structures are transported at speeds of 1–4 μm per second.

The nuclear envelope is composed of two membranes, each of which is a lipid bilayer. Within the nuclear envelope are nuclear pores, which number between 3000 and 5000 in a typical growing cell. The nuclear pores act as gatekeepers, controlling the transport of materials in and out of the nucleus. Traffic into the nucleus includes ribosomal proteins and histones, while traffic out of the nucleus includes ribosomes, mRNAs and transfer RNAs.

Each nuclear pore is constructed from a complex of more than 50 protein species that control the movement of materials in and out of the nucleus. The protein molecules are assembled into discrete structures, including a 'basket' protruding into the nucleus and eight filaments that protrude into the cytoplasm. At the centre of the pore

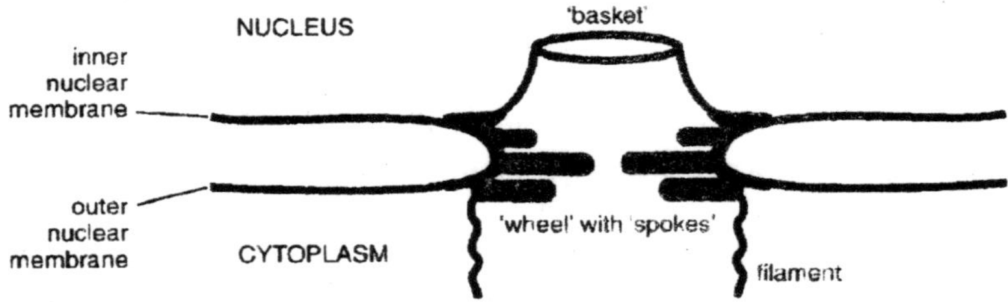

Fig. 9.3. Structure of the nuclear envelope and a nuclear pore.

is a channel through which molecules can diffuse, though most molecules and particles pass through the pore complexed with specific carrier proteins (importins and exportins). The channel can open to allow the passage of particles up to 25 nm across, and there have been reports of larger particles being moved through nuclear pores. Small nucleocapsids/virions, such as the parvovirus virion, can pass through, but larger viruses must either shed some of their load to form slimmer structures or uncoat at a nuclear pore.

Genome Uncoating

Uncoating can be defined as the complete or partial removal of the capsid to release the virus genome. Depending on the virus, the process can take place

1. at the cell surface, the capsid remaining on the exterior surface of the cell;
2. within the cytoplasm;
3. at a nuclear pore;
4. within the nucleus.

It should be noted that successful entry of a virion into a cell is not always followed by virus replication. The host's intracellular defences, such as lysosomal enzymes, may inactivate infectivity before or after uncoating. In some cases the virus genome may initiate a latent infection rather than a complete replication cycle.

In at least some cases, however, provided that the uncoated virus genome survives intact, and is at the correct location in the cell, then its transcription (or in some cases translation) can begin.

Bacteriophages

Like animal viruses, phages bind specifically to cell surface molecules that function as receptors and co-receptors. For many phages these molecules are on the surface of the host cell wall, but for some phages they are on the surface of other structures (pili, flagella or capsules). Many of the virus attachment sites are on particular virion structures, such as the tail fibres of phage T4. As with animal viruses, the attachment of a phage to its host is initially reversible, and becomes irreversible after attachment to further receptors and/or co-receptors.

Infection of a cell with a phage involves entry into the cell of the virus genome and perhaps a few associated proteins, the capsid and any associated appendages remaining at the cell surface. This is in contrast to most animal and plant viruses, where the entire virion, or at least the nucleocapsid, enters the host cell. Unless the host is a mycoplasma (mycoplasmas do not have cell walls), delivery of a phage genome into a cell requires penetration of a cell wall and perhaps also a slime layer or capsule; the virions of many phages carry enzymes, such as lysozymes, to aid this process.

Survival of Infectivity

Preservation of Virus Infectivity

There are several practical situations in which there is a need to preserve virus infectivity. A prime example is during the storage and distribution of live virus vaccines, when there is a need to retain the potency of the vaccine. A second example is during the transport and storage of blood, faeces, saliva and other specimens from human and animal patients prior to testing in a diagnostic virology laboratory. Last, but by no means least, virologists need to preserve the infectivity of the viruses that they work with.

Destruction of Virus Infectivity

There are many situations in which there is a requirement to destroy virus infectivity. Procedures to minimize the transmission of pathogenic viruses (and other micro-organisms) in hospitals and other health care facilities include the disinfection of surfaces and the sterilization of equipment and materials. A good example of a situation requiring rigorous action is an outbreak of winter vomiting disease caused by a norovirus; patients' vomit contains significant quantities of virus, which must be inactivated if the outbreak is to be brought under control.

Inactivated (killed) virus vaccines are made from preparations of virulent viruses, so it is essential that all infectivity is destroyed. Blood products, such as clotting factors for haemophiliacs and immunoglobulins, are treated to destroy any viruses (e.g. HIV, hepatitis viruses) that might be present in the donated blood. Continuing the theme of human and animal health, water supplies that are free from pathogenic viruses (and other micro-organisms) are of fundamental importance, so water treatment procedures must include processes that destroy these agents. The same applies to the treatment of swimming pool water.

The processes for production of cheese, yogurt and other dairy products utilize lactic acid bacteria to ferment the milk. There are many phages that infect these bacteria, and a phage infection during a production process can result in poor product quality, or in delay or failure of the fermentation. To minimize the risk, procedures are required for treatment of the milk or whey to inactivate any phages present, and for the disinfection of the factory environment.

Finally, virologists need supplies of pipettes, cell culture vessels and other laboratory materials that are guaranteed free from viruses (and other microorganisms), then after these materials have been used any viruses present must be inactivated before disposal.

Inactivation Targets in Virions

The infectivity of a virion can be destroyed by the alteration or removal of its nucleic acid and/or one or more of its proteins. A protein molecule may be altered by the induction of a conformational change or, more drastically, by the breaking of covalent bonds, such as peptide bonds and disulphide bonds.

Inactivation Kinetics

The rate at which a virus loses infectivity under defined conditions (for example, at a particular temperature and pH value) can be determined by incubating aliquots of the virus preparation under those conditions. At various time intervals the surviving infectivity in an aliquot is determined in an infectivity assay appropriate for the virus. The logarithm of the percentage infectivity surviving is plotted against time. In some circumstances infectivity is lost at a constant rate until no further infectivity can be detected. The slope of the line is a measure of the rate of infectivity loss. Sometimes there is a change in the rate of infectivity loss after most (usually more than 99 per cent) of the infectivity has been destroyed. It is not entirely clear why a small fraction of some virus preparations loses infectivity at a slower rate. It has been suggested that this fraction might contain virions that have a higher degree of resistance to the inactivating factor and/or that the preparation might contain virion clumps, with virions at the centres of the clumps having some protection.

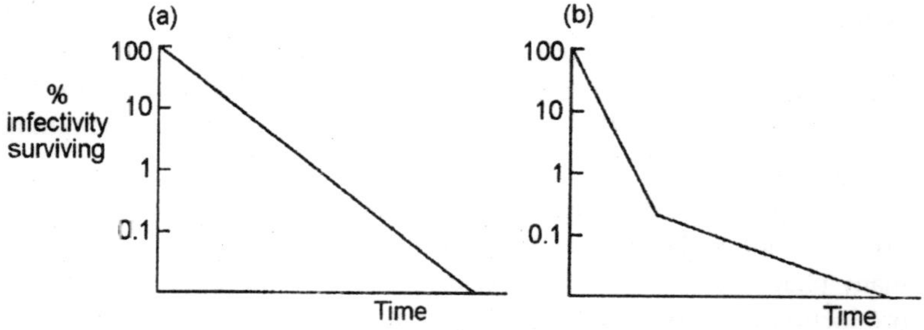

Fig. 9.4. Loss of virus infectivity at one rate (a) and at two rates (b).

Agents that Inactivate Virus Infectivity

Physical Agents

Temperature

All infective viruses lose infectivity over time, the rate of infectivity loss being highly dependent on temperature. There is a lot of variability between viruses in the rates at which they are inactivated. In general, naked viruses are more stable than enveloped viruses; most naked viruses lose little infectivity after several hours at 37°C, whereas most enveloped viruses lose significant infectivity under the same conditions. The generalization is, as stated, a broad one, as there may be significant differences in stability between closely related viruses. Under dry conditions varicellazoster virus survives well, while the related herpes simplex viruses survive poorly.

At temperatures below —35°C most viruses lose infectivity very slowly, so for long-term storage of viruses they are placed in freezers or in liquid nitrogen. Preservation of infectivity may be enhanced by storing virus preparations in a freeze-dried form, rather than as aqueous suspensions. Moving to higher temperatures, most viruses are completely inactivated within 30 minutes at 60°C or within a few seconds at 100 °C. Some of the bacteriophages of lactic acid bacteria that cause problems in the dairy industry are quite heat resistant, and some infectivity may survive the heat treatments used in the decontamination procedures.

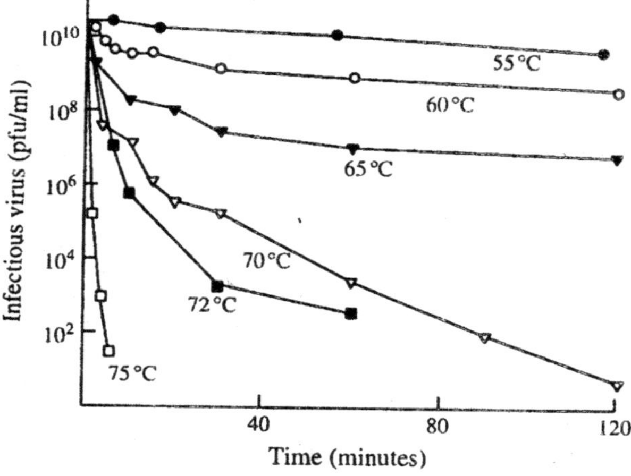

Fig. 9.5. Inactivation of Lactococcus lactis phage P008 infectivity in a broth at temperatures between 55 and 75°C.

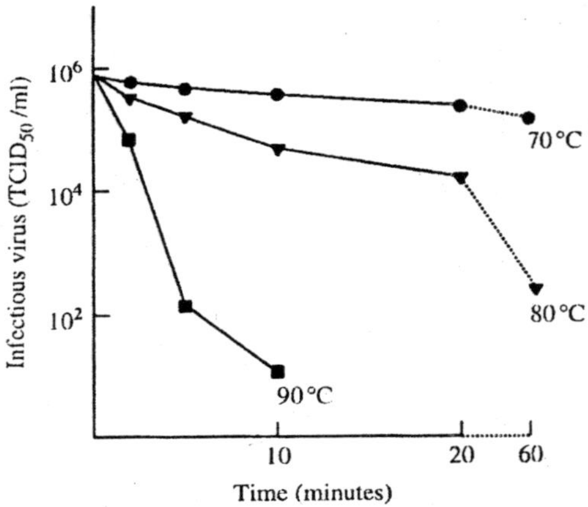

Fig. 9.6. Inactivation of minute virus of mice infectivity in water at 70, 80, and 90°C.

The virions of some viruses, such as hepatitis A virus (HAV) and parvoviruses, are exceptionally heat stable. Data for the minute virus of mice (a parvovirus) indicate some infectivity surviving after 60 minutes at 80°C.

Exceptions to the generalization concerning the stability of naked and enveloped virions include those of hepatitis B virus (HBV), which are enveloped but have high stability, and those of poliovirus, which are naked but have low stability. Interestingly, HAV (heat-stable virions) and poliovirus (heat-labile virions) are both members of the family *Picornaviridae*, so, as with the herpesviruses, there is not always consistency in the degree of stability between viruses with similar structures.

The fact that virus preparations lose infectivity means that special precautions have to be taken to minimize infectivity loss during transport and storage of live vaccines, such as polio, mumps, measles and rubella. These precautions include a 'cold chain' of containers and vehicles that are refrigerated and/or insulated. This adds considerably to the expense of vaccination programmes.

The rate of infectivity loss depends on the nature of the medium containing the virus. The presence of proteins, fats, glycerol or certain salts may protect infectivity. Magnesium ions reduce the rate of inactivation of many RNA viruses, including poliovirus, so magnesium chloride is added to live polio vaccines to extend their shelf life.

The most heat-resistant viruses known are those that infect the thermophilic prokaryotes that inhabit hot springs. The optimum temperature for the growth of many of the hosts of these viruses is around 85°C. Prions are also remarkably heat resistant.

Radiation

Some forms of radiation inactivate virus infectivity because they damage nucleic acids; for example, X rays, gamma rays and ultraviolet rays cause breaks in nucleic acids. Ultraviolet rays also cause other types of damage to nucleic acids, including the formation of thymine dimers in dsDNA. The thymine-adenine hydrogen bonds at two adjacent base pairs are broken and covalent bonds are formed between the two thymine residues.

Sunlight inactivates virus infectivity because of its ultraviolet component. A practical aspect of this is the rapid inactivation by sunlight of baculoviruses applied to crops and forests as insecticides.

Chemical Agents

Chemicals that inactivate virus infectivity do so with similar kinetics to physical agents, with either one or two inactivation rates. When some inactivation data are plotted the initial part of the graph has a small shoulder before the straight line. This could be due to a delay while the chemical reaches its target in the virion. Many chemicals can inactivate virus infectivity, but for many of them the inactivation target(s) are uncertain. We shall consider the effects of hydrogen ion concentration (pH) and of a variety of other chemical agents on virus infectivity.

Acids and alkalis

Most viruses are fairly stable at pH values between 5 and 9, but at more extreme pH values virions may undergo conformational changes and rates of infectivity loss usually increase. At high pH values both nucleic acids and proteins can be damaged. Alkali, in the form of dilute sodium hydroxide (sometimes in combination with steam), is used to disinfect manufacturing vessels and tools in the pharmaceutical industry. The infectivity of some viruses is inactivated at low pH values as a result of alterations to surface proteins. Viruses that infect humans and other mammals via the gastrointestinal route, however, are relatively stable at low pH values. These viruses, which include rotaviruses, have to survive in the stomach contents, which are commonly at a pH value between 1 and 2. Furthermore, their virions must be able to withstand the sudden increase in pH when they are moved from the stomach to the intestine (pH about 7.5), and they must be resistant to inactivation by bile salts and proteolytic enzymes.

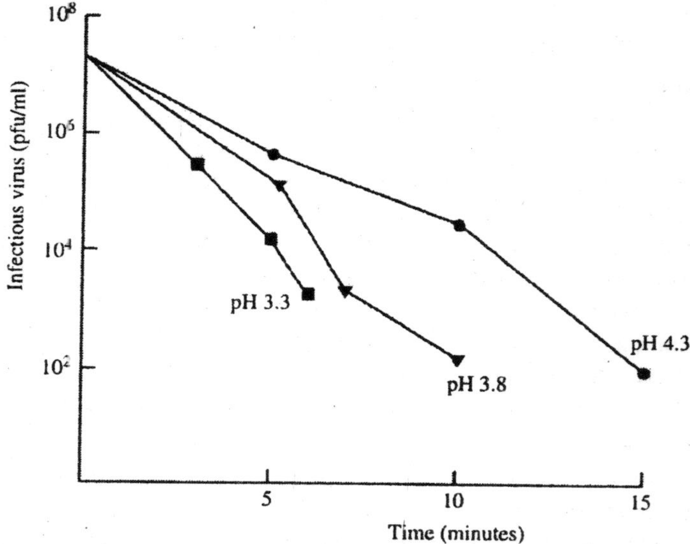

Fig. 9.7. Inactivation of hepatitis a virus infectivity at three pH values at 85°C.

The enteroviruses and HAV infect their hosts via the gastro-intestinal tract. These viruses are picornaviruses, but, when compared to those picornaviruses (such as rhinoviruses and aphthoviruses) that infect their hosts via other routes, their virions are much more acid resistant. It can be seen that HAV infectivity can still be detected after 6 minutes at pH 3.3 at a temperature of 85°C, thus HAV is an example of a virus with high resistance to inactivation by both heat and acid.

Phenol, which is a weak acid, destroys the capsids of virions without altering nucleic acids. It is commonly used in the laboratory to extract nucleic acids from virions.

Hypochlorite

One of the most widely used disinfectants is hypochlorite. Both the hypochlorite ion (OCl^-) and the undissociated hypochlorous acid (HOCl) are excellent disinfectants; they act by oxidizing organic molecules, such as proteins. Hypochlorite ions are either supplied as sodium hypochlorite or they are generated by dissolving chlorine gas in water. The latter process is used to treat drinking water supplies and swimming pool water.

Sodium hypochlorite is one of the disinfectants used in the dairy industry to control phages of lactic acid bacteria such as *Lactobacillus*

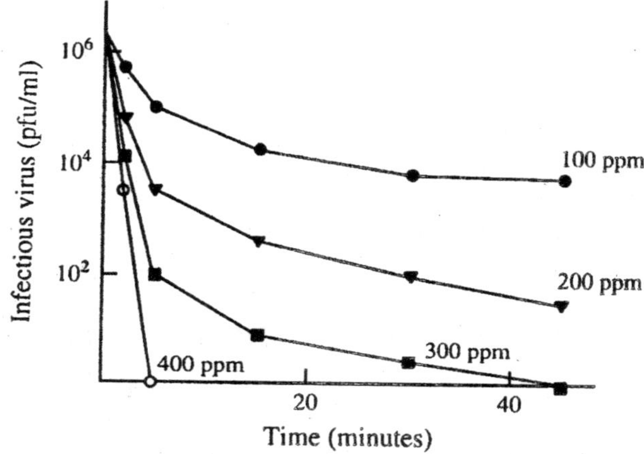

Fig. 9.8. Inactivation of Lactobacillus phage BYM by hypochlorite.

spp. The inactivation of a *Lactobacillus* phage at four concentrations of hypochlorite is shown in Figure 9.8. At the three highest concentrations of hypochlorite there were slower rates of inactivation for a small percentage of the virions.

Aldehydes and alcohols

Formaldehyde and glutaraldehyde damage both enveloped and non-enveloped virions. Formaldehyde cross-links proteins and reacts with amino groups in nucleic acids; it is used to destroy the infectivity of poliovirus and other viruses for use in inactivated vaccines. Glutaraldehyde is useful for disinfecting metal surfaces because it is noncorrosive. Ethanol and isopropyl alcohol are used as general purpose disinfectants and in waterless hand washes.

Lipid solvents

Detergents and many organic solvents, such as chloroform, destroy the infectivity of enveloped viruses by removing their lipid membranes and associated glycoproteins. Detergents are used in the preparation of clotting factors and other blood products to destroy enveloped viruses (particularly HIV, HBV and HCV) that may be present in pooled blood donations. Lipid solvents may also damage virions with internal lipid, such as the corticoviruses and the tectiviruses. Quaternary ammonium compounds are cationic and disrupt virion envelopes. A number are used as disinfectants, especially chlorhexidine gluconate, which is widely used in health care institutions. Anionic detergents, such as sodium dodecyl sulphate, disrupt virion envelopes and capsids.

10

Antiviral Susceptibility Testing

Antiviral therapy for the successful management of viral infections has come of age. Within the last decade, rapid advances have been made in the development of new antiviral agents. Many of these drugs are now available for the treatment and/or prevention of infections caused by herpes simplex virus (HSV), varicella-zoster virus (VZV), cytomegalovirus (CMV), the human immunodeficiency virus (HIV), influenza A virus, respiratory syncytial virus (RSV), and hepatits B and C viruses (HBV, HCV). An important caveat to the increased number and usage of antiviral agents has been the emergence of drug-resistant viruses, particularly in patients with acquired immunodeficiency syndrome (AIDS) or other conditions of immunosuppression such as hemotologic malignacy and organ transplantation. Although still uncommon, drug resistance has been reported after therapy with a number of antiviral agents. As more patients fail to respond to appropriate therapy and additional antiviral agents are produced, a definite need arises for the diagnostic laboratory to provide rapid and practical antiviral susceptibiltiy testing to assist the physician in defining drug resistance and choosing appropriate alternative therapies.

Numerous procedures have been described for testing the susceptibility of viruses to antiviral agents. In general, all antiviral susceptibility assays require viral replication in a suitable host cell. However, each assay employs a different method for measuring the activity of an antiviral agent against the replication of the virus. The inhibition of virus-induced cytopathic effect (CPE) or plaque formation;

a decrease in the production of viral antigens, enzyme activities, or total virus yield; a reduction in viral nucleic acid synthesis; and the inhibition of cell transformation are parameters that can be monitored in an *in vitro* system. This chapter focuses on the use of a newly developed DNA-DNA hybridization assay as it applies to susceptibility testing of HSV isolates.

Acyclovir has been widely used as an effective treatment for and prophylactic against HSV infections. However, resistance of HSV to acyclovir has emerged with the occurrence of chronic, progressive, debilitating disease in immunocompromised patients receiving prolonged or multiple courses of therapy. Resistance to acyclovir can result from mutations in the genes encoding the viral thymidine kinase or the viral DNA polymerase. Most acyclovir-resistant HSV isolates are deficient in, or have altered production of, viral thymidine kinase activity. The frequency with which resistant strains are being found and the increasing morbidity and mortality associated with these virus strains is a problem of growing concern. Foscarnet has been employed successfully as an alternative antiviral agent for treating thymidine kinase-deficient or -altered acyclovir-resistant HSV. However, resistance to this drug has also been documented.

Two commonly employed methods for antiviral susceptibility testing of HSV isolates are the plaque reduction and dye uptake assays. The plaque reduction assay has long been considered the standard method used to determine the suceptibiltiy of all cytopathogenic viruses to antiviral agents. In this procedure, cell monolayers are established in 12-well tissue culture plates, infected with 100-200 plaque-forming units (PFU) of virus, and overlaid with medium containing various concentrations of drug and 0.4% agarose, 1-2% methylcellulose, or serum immunoglobulin. Following a 72-hr incubation, the plates are fixed and stained with crystal violet dye. The plaques are then counted; the concentration of drug causing a 50% inhibition in plaque formation represents the 50% inhibitory dose (ID_{50}) for the antiviral agent. The plaque reduction assay is accurate, reliable, and relatively simple to perform and does not require specialized equipment. It is difficult to automate, however, and is best suited to testing small numbers of specimens. A major disadvantage of this assay is that the enumeration of plaques is tedious, subjective, and time consuming and the number of plaques that can be counted accurately is limited. The infectivity titer of the virus must also be determined before the assay is performed, leading to a delay in reporting results. The dye uptake assay is a colorimetric method that involves the quantification of neutral red dye that is preferentially taken up by viable host cells following incubation with virus and various concentrations of antiviral drug. The assay is semi-automated and performed in a 96-well microtiter plate. The ID_{50} value is the concentration of drug causing a 50% reduction in viral CPE compared to cell and virus controls. The assay is reproducible and provides an objective end point. It can be performed on large numbers of clinical isolates and can be adapted for use with different drugs and viruses. However, the infectivity titer of each virus isolate must be determined, and the cost of automating the assay is a factor. The test can be performed manually, but is cumbersome and slow.

A DNA—DNA hybridization assay has been developed that measures the reduction in viral nucleic acid synthesis when virus-infected cells are incubated in the presence of antiviral agent. The assay is simple to perform, is less labor intensive than the conventional methods, and does not require the determination of virus infectivity titers before performing the assay. Like the dye uptake assay, the DNA—DNA hybridization system can be expanded to accommodate multiple clinical isolates and antiviral drugs. The procedure is rapid, providing a reliable determination of the susceptibiltiy of virus isolates to antiviral agents within 48 hrs. AlthoUgh the hybridization assay has the disadvantage of using a radioiodinated probe, it can be purchased as a ready-to-use kit that contain cells, medium, and reagents (except antiviral drug) required to perform the test. Formated kits are currently available for HSV, CMV, and VZV.

Principle of the Test

Radiometric DNA—DNA hybridization is used to measure the effects of antiviral agents on the replication of HSV type 1 or 2. Cultured cells in 24- well plates are infected with a fixed inoculum of virus and overlayed with tissue culture medium containing various concentrations of an antiviral agent. Following incubation at 37°C for 24-48 hr, the infected cells from each well are lysed and the DNA is denatured and immobilized on separate filter membrane supports or wicks. The bound DNA is hybridized to an HSV-specific, ^{125}I-labeled probe and the amount of HSV target DNA from each well is determined by measuring the radioactivity in a gamma counter. The ID_{50} for the antiviral agent is calculated using linear regression and is defined as the concentration of drug that results in a 50% reduction in DNA hybridization relative to virus-infected control wells containing no drug. The resistance or susceptibility of a particular HSV isolate is established by using an appropriate end point for the antiviral agent studied.

Collection and Preparation of Herpes Simplex Virus Isolates

To perform the antiviral susceptibility assay, a viable isolate of HSV is required. If necessary, clinical specimens should be collected from dermal, genital, oral–facial, or other appropriate body sites and submitted to a diagnostic laboratory for virus isolation. Otherwise, clinical isolates of HSV can be sent directly to the laboratory for testing. These samples can be submitted as frozen isolates stored at -70°C or as freshly isolated virus demonstrating at least 50% CPE in cultured cells. When submitting fresh isolates, the infected tissue culture tube should be filled completely with an appropriate medium such as modified Eagle's minimum essential medium (EMEM) supplemented with 2 mM glutamine and 2% fetal bovine serum (FBS). The cap should be firmly tightened and covered with parafilm, and the tube should be transported at ambient temperature to the laboratory. Once received in the laboratory, all isolates should be routinely passaged once and regrown to 50-100% CPE in culture tubes of the same cell type. A number of cell types can be used for this purpose including primary rabbit kidney, primary human nenonatal kidney, mink lung, continuous African Green monkey kidney (CV-1 or Vero), continuous human lung carcinoma (A-549), or human diploid fibroblast (MRC-5 or WI-38) cells. CV-1 cells are preferred because this cell line is used to perform the antiviral susceptibility assay. All HSV isolates should be at the first or second passage level; continuous subpassage should be avoided prior to performing susceptibility testing. Aliquots of each isolate should be harvested and frozen at -70°C for future use.

Preliminary Considerations

Materials

Media and reagents

Hybriwix Herpes (HSV) Antiviral Susceptibility Test Kit (Diagnostic Hybrids, Athens, OH):

1. 3 24-well cell culture dishes containing monolayers of CV-1 cells in EMEM with 10% FBS and gentamicin.
2. 3 27-ml bottles sterile replacement medium containing EMEM with 5% FBS, gentamicin, and amphotericin B.

3. 6 ml DNA wicking agent (lysis solution) containing 0.5 *N* NaOH and sodium dodecyl sulfate (SDS) in a proprietary formulation; this reagent is used for the single-step lysis of cell monolayers and denaturation of released viral DNA.
4. 36 Hybriwix combs, each containing two 9 × 20-mm consecutively numbered wicks.
5. 6 positive and 9 negative control wicks; the negative control wicks are pretreated with a blocking reagent to simulate a negative reaction; purified pSP64 plasmid DNA containing HSV sequences complementary to the HSV DNA probe is absorbed to the positive control wicks.
6. 60 ml wash reagent containing a 20× concentration of 0.3 *M* sodium chloride, 0.03 *M* sodium citrate, and 2% SDS in distilled water.
7. 3-3.5-ml vials of HSV probe hybridization agent containing <1.6 μCi ^{125}I-labeled HSV DNA in 50% formamide, 0.5 *M* NaC1, 0.05 *M* sodium citrate, and distilled water; the probe is group specific and is produced from a clone of HSV-1 F DNA in bacteriophage M13 mp19.
8. Wash container.
9. Radioactive symbol label.

- Stock solutions of acyclovir [9-(2-hydroxyethoxymethyl)guanine] and foscarnet (trisodium phosphonoformate), prepared in sterile distilled water at concentrations that are convenient for preparing a range of drug dilutions to be used in the susceptibility assay; final stock concentrations of 640 μg/ml acyclovir and 1800 μg/ml foscarnet are suggested; the concentrations of stock solutions may vary depending on the range of drug concentrations selected for testing and on the antiviral agent employed
- EMEM with Earle's balanced salt solution, supplemented with 2 mM L-glutamine, 20 mM HEPES, and either 10% (cell growth medium) or 2% (cell maintenance medium) heat-inactivated (56°C, 30 min) FBS
- CV-1 cells, propagated in the laboratory, as an alternative to purchasing them as part of the commercial susceptibility test kit; stock flasks of these cells can be maintained and used to prepare monolayers in tissue culture tubes and 24-well plates.
- Fresh or frozen virus isolates, including acyclovir and foscarnet-sensitive and -resistant control strains of HSV
- Distilled water

Supplies

1. Sterile 1.0-, 5.0-, 10.0-, and 25-ml serological pipettes; disposable plastic pipettes are commercially available and convenient to use
2. Sterile unplugged pasteur pipettes
3. Sterile 15-ml and 50-ml polypropylene centrifuge tubes for dilution of HSV isolates and antiviral agents
4. Culture plates, 24-well flat-bottom
5. Tissue culture flasks, 80 cm^2
6. 16 × 125-mm Tissue culture tubes
7. Forceps
8. Cryovials for freezing HSV isolates and stock solutions of antiviral agents
9. 0.5% Hypochlorite solution for cleaning work surfaces
10. Racks designed to hold tissue culture tubes at a 5-7° angle
11. Safety pipetting devices, protective clothing, latex gloves, infectious-waste disposal containers, and adequate sterilization facilities
12. Linear regression probability program for calculation of ID_{50} values
13. Scissors
14. 12 × 55-mm Gamma counting vials (Sarstedt)
15. ^{129}I reference standard

Equipment

1. Inverted and standard light microscopes capable of 100× to 200× magnification
2. Class II biological safety cabinet
3. Refrigerator at 2-8°C
4. Freezer at –70°C
5. Vacuum source and collection trap containing 0.5% sodium hypochlorite to be used for collecting aspirated infectious waste
6. Vacuum source and collection trap suitable for collecting low-level radioactive liquid waste
7. Humidified incubator at 35-37°C with 5% CO_2
8. Gamma counter
9. Water baths at 56°C for heat-inactivation of FBS, and 60°C and 73°C for hybridization assay

Storage Instructions

1. On receipt of the Hybriwix Herpes (HSV) Antiviral Susceptibility Test Kit, carefully remove the aluminum foil cover from all cell

culture dishes, place the plastic lids on the dishes, and incubate at 35-37°C in a humidified atmosphere of 5% CO_2 for 48 hr or until the cell monolayers are confluent. The sterile replacement medium should be stored at 2-8°C and all other reagents and kit accessories should be stored at room temperature. The radioactive probe should be stored in the original container in a specially designated area of the laboratory in accordance with federal and state regulations. The shelf-life of the kit is 2 wks.

2. The acyclovir and foscarnet stock solutions should be stored in single- use aliquots of 1.5 ml at −70°C no longer than 1 yr. The antiviral agents should be used immediately after thawing. Repeated freezing and thawing or storage under conditions other than those recommended should be avoided.
3. EMEM with L-glutamine and FBS should be stored at 4°C no longer than 1 mo.
4. Expiration dates should be indicated on the labels of all media and reagents. The expiration date established by the manufacturer should be observed.

Preliminary Preparations

Drug dilutions

1. The dilutions and volumes of drug needed to perform a susceptibility test are dependent on the range of drug concentrations employed and the total number of HSV isolates to be tested. **Figures 3** and **4** illustrate the typical ranges of acyclovir and foscarnet, respectively, employed in the HSV hybridization assay and the configurations of the 24-well plates when one clinical HSV isolate and two control virus strains are examined.
2. Sufficient volumes of individual drug dilutions should be prepared in sterile EMEM containing 5% FBS and 2 mM L-glutamine (replacement medium) so 1.0 ml of each drug concentration can be added to the appropriate wells of a 24-well plate.
3. All drug dilutions should be prepared immediately before use and each dilution should be mixed thoroughly before making the next dilution.

Dilution of clinical HSV isolates and control strains

The optimum virus inoculum for the susceptibility assay can range from 100 to 10,000 PFU per well. This level can be achieved by growing a clinical HSV isolate or control virus strain to 50-100% CPE in a single 16× 125-mm tissue culture tube of CV-1 cells. Virus

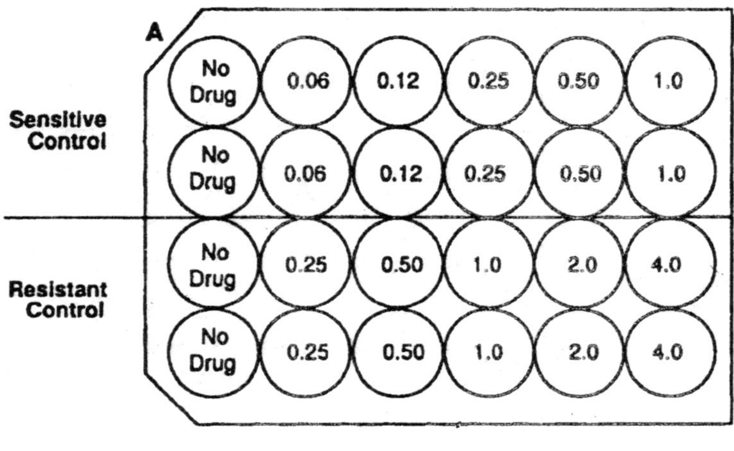

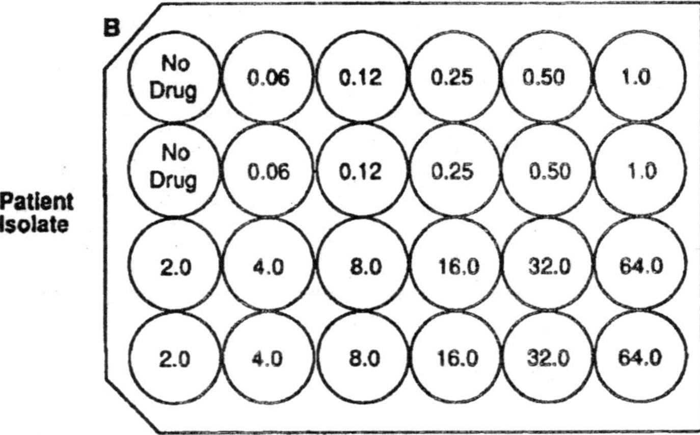

Fig. 10.1. Illustration of the typical ranges of acyclovir (μg/ml) employed in the HSV hybridization assay, and the configuration of the 24-well plates when two control virus strains (A) and one clinical HSV isolate (B) are examined.

isolates are usually harvested when approximately 75% of the monolayer is infected.

1. Scrape the infected cell monolayer from the tube surface into 2.0 ml culture medium using the tip of a sterile disposable 1.0-ml pipette.
2. Disperse the cells by gently pipetting them up and down enough times to break apart cell aggregates and produce an even suspension.

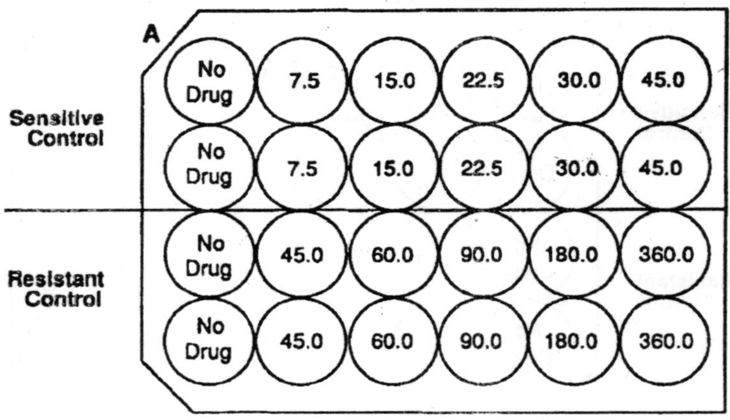

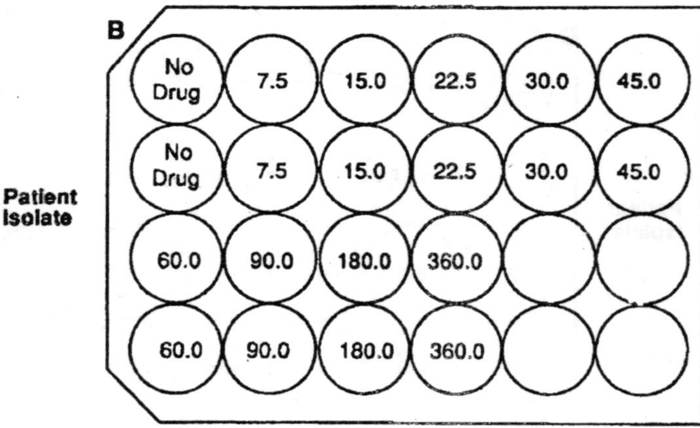

Fig. 10.2. Illustration of the typical ranges of foscarnet (µg/ml) employed in the HSV hybridization assay and the configuration of the 24-well plates when two control virus strains (A) and one clinical isolate (B) are examined.

3. Dilute this cell suspension 1000-fold with EMEM containing 10% FBS and 2 mM L-glutamine. Inoculate 0.2 ml into each well of 24-well plate.

Using this dilution scheme, an inoculum containing approximately 2,000-3,000 PFU is obtained that is suitable for optimum results.

24-Well cell culture plates

1. Determine the number of 24-well plates of CV-1 cells needed to perform a given susceptibility assay and order the appropriate number of test kits to be used.

2. As a general guide, for each antiviral agent employed, a single 24- well plate is needed for each clinical HSV isolate and an additional plate is required for the control strains. Therefore, a total of two clinical isolates and two controls can be tested using the described format and a single 72-assay Hybriwix kit.
3. The kits may be purchased at a lower cost without cultured CV-1 cells, but the cells must then be maintained and propagated in the laboratory for preparation of 24-well plates.
4. The cell monolayers of the plates should be examined by microscopy before use to determine the quality of the cells. At the time of virus inoculation, the cell monolayers should be freshly confluent and not overgrown.

Preliminary Comments and Precautions

1. Bring all reagents and media to room temperature (20-25°C) before use and return them to appropriate storage temperatures immediately after use.
2. Do not smoke, eat, or drink in areas where specimens, virus isolates, or kit reagents are handled.
3. Take appropriate precautions when handling infectious and radioactive materials. Do not mouth pipette samples or reagents. Avoid contact with broken skin or mucous membranes. Wear disposable gloves, laboratory coats, and other appropriate protective devices. Wash hands thoroughly after handling these materials.
4. Use radioactive material only in designated work areas and cover laboratory bench surfaces with an absorbent material.
5. Use an absorbent material and suitable detergent to clean radioactive spills from involved surfaces and 0.5% sodium hypochlorite to wipe all areas before and after processing of infectious material. Dispose of all infectious and radioactive materials properly.
6. Do not use materials and reagents beyond the assigned expiration dates and do not interchange or mix different lots of reagents.
7. Arrange materials and equipment to provide easy access and to minimize the number of manipulations.
8. Perform accurate pipetting and careful dilution of antiviral drugs and virus isolates for quality results using this procedure. Use separate sterile pipettes when preparing each dilution and during the inoculation of each virus isolate.
9. Use appropriate aseptic technique throughout the procedure.

10. Avoid contact of DNA wicking agent with skin and clothing, since it contains dilute sodium hydroxide (NaOH).
11. Use forceps to manipulate processed Hybriwix combs and wicks.

Antiviral Susceptibility Assay

Virus Inoculation and Drug Addition

1. Obtain a culture tube of HSV-infected CV-1 cells demonstrating 50-100% CPE.
2. Prepare a cell suspension and dilute the infected cells as described above.
3. Remove a 24-well plate of freshly confluent CV-1 cells from the incubator and gently aspirate the medium from each of the 24 wells, taking care not to disturb the cell monolayer.
4. Inoculate the wells with 0.2 ml diluted HSV suspension. Clearly label the plate and wells with an appropriate identification number.
5. Incubate the plate for 60-90 min at 35-37°C in a humidified incubator with 5% CO_2 to allow for virus adsorption. Gently rock the plate back and forth every 15 min to prevent drying of the monolayer.
6. After virus adsorption, gently aspirate the virus inoculum from each well and overlay duplicate wells with 1.0 ml each of the diluted antiviral agent. Two wells should receive 1.0 ml replacement medium to which no drug has been added. These wells serve as untreated controls.
7. Incubate the plate at 35-37°C in the humidified CO_2 incubator for 24-48 hr. Best results are obtained when the CPE in the "no drug" control wells reaches 50-100% within this time period. When the virus inoculum is prepared as described earlier, the inoculated plates normally require 48 hr to achieve this level of CPE. The monolayers should be examined daily for progression of viral CPE and possible deterioration of the cell monolayers.

DNA–DNA Hybridization

1. Following the 48-hr incubation, gently but completely aspirate the medium from each well of the 24-well plate. Take care not to disturb the cell monolayers.
2. Immediately add two drops of DNA wicking agent (lysis solution) to each well to lyse the cells and virus particles and to denature the DNA. Rock and tap the plate with enough force to ensure that the monolayer in each well is completely covered with the lysis buffer.

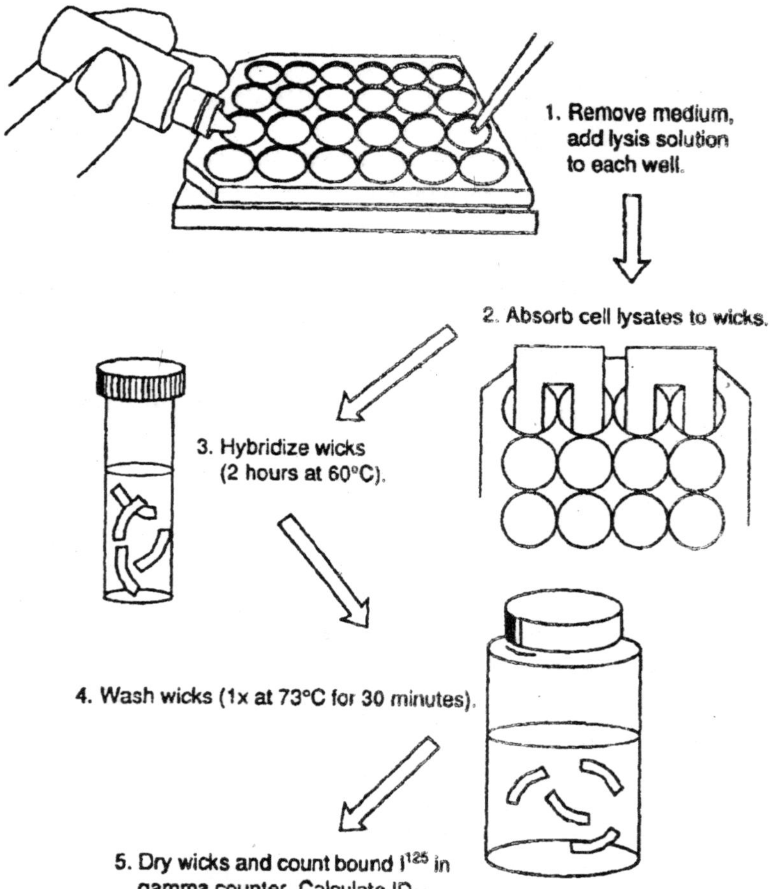

Fig. 10.3. DNA-DNA hybridization assay for measuring the susceptibility of HSV to antiviral agents.

3. Using forceps, arrange the Hybriwix combs to be used in numerical order on a clean surface or paper towel.
4. Carefully insert the wicks of the combs into their respective duplicate wells and allow the cell lysates to absorb vertically by capillary movement (wicking) onto the wicks. Format the wicks so that increasing wick numbers will correspond to increasing drug concentrations. It is important to place the wicks in the wells so that the DNA from the lysed cells moves upward from the bottom of the wick. The wicks should rest diagonally in the

wells and should not stick to the sides of the wells. Wicking is complete when no DNA wicking agent is left in the wells and the Hybriwix comb is completely wet. The DNA should be in the lower third of the wick.

5. Once wicking is completed, use forceps to remove the combs from the wells and place them on a clean paper towel. Using scissors, cut the wicks from the combs.
6. Transfer the wicks to a vial of HSV probe hybridization agent (^{125}I- labeled HSV group-specific DNA probe), using one vial for every 24 wicks. Include three negative and two positive control wicks in each hybridization vial.
7. Incubate the hybridization vials for 2 hr at 60 ± 2°C. Also place 114 ml distilled water at 73°C in preparation for Step 12.
8. After hybridization, carefully aspirate the probe hybridization agent from each vial into a trap suitable for collecting low-level radioactive liquid waste.
9. Using a pipette, add 4.0 ml distilled water to cover the wicks in the hybridization vials. Cap the vials and invert several times to rinse the contents.
10. Aspirate this rinse solution into a trap suitable for the collection of radioactive liquid waste.
11. Add 6.0 ml wash reagent to the provided wash container.
12. Carefully transfer the wicks from the vials to the wash container and gently swirl. At this point, add 114 ml preheated (73°C) distilled water to the wash container.
13. Place the wash container in a 73°C water bath for 30 min.
14. Following the 30 min incubation, aspirate the wash into a trap suitable for the collection of radioactive liquid waste.
15. Using forceps, separate the wicks and arrange them in numerical order on a clean paper towel.
16. Transfer the wicks to appropriate vials and count each wick for 2 min in a gamma counter set for a ^{125}I label. Make certain each wick is at the bottom of the vial and that the window of the gamma counter is set to count the entire surface of a wick. The amount of radioactivity present on each wick is dependent on the amount of HSV DNA present after culture in the presence of a specific antiviral agent.
17. Perform appropriate controls for each assay. Calculate the 50% inhibitory dose (ID_{50}) for each drug and virus isolate tested.

Procedural Notes

1. Care must be taken to use an appropriate virus inoculum in the susceptibility assay. As previously stated, optimum results are achieved when the CPE in the "no drug" control wells is approximately 75% (50-100%) at the end of the incubation period. If a shorter time is required to reach this level of CPE, the HSV inoculum may be too great and the virus may kill the cells before the antiviral agent can be assimilated and have an effect. The opposite effect may occur if the virus inoculum is too weak and too much time is required to achieve the appropriate CPE. The antiviral agent may overwhelm the virus isolate, giving a false impression of the effectiveness of the drug.
2. When aspirating culture medium and reagents from the wells of the cell culture plates, use a sterile disposable pasteur pipette attached to the tubing from a vacuum aspirator and trap. The vacuum should yield gentle aspiration without disturbing the cell monolayer. Hold the 24-well plate at a 20-30° angle and aspirate the medium at the meniscus, directing the tip of the pipette at a 45° angle and moving down the side of each well to the bottom. Completely aspirate all residual material from the wells.
3. Approximately 15 μl DNA wicking agent is used to lyse cells and virus particles within each well. Do not allow this volume to evaporate before placement of the wicks into the wells. By working at a steady rate, multiple plates can be lysed and wicked without a loss in volume of the lysis solution.
4. Strictly adhere to incubation temperatures and times to avoid false results.
5. If probe hybridization cannot be performed immediately after lysis and wicking, allow the wicks to air dry and store them at room temperature. The wicks should be hybridized within 72 hr.
6. The DNA wicking reagent, the 20× wash reagent, and the HSV probe hybridization agent may become turbid or form a precipitate at temperatures below 20°C. If this occurs, warm the reagents in a 37°C water bath until the precipitate dissolves and the solution becomes clear. Precipitation of these reagents will affect the performance of the assay, so close inspection of the solutions before use is important.
7. Instability, deterioration, or toxicity of tissue culture medium, cell monolayers, antiviral agents, and assay reagents should be monitored continuously. Observe medium for microbial contamination; the

formation of precipitates, crystals, or sediments; and the ability to support cell growth and viral infection. Determine whether the antiviral agents are cytotoxic to the CV-1 cells. To do this, examine the effect the various concentrations of drug have on normal cell morphology when the drug is incubated with uninfected cells for 48 hr. Define the dose(s) of each antiviral agent required to cause a microscopically visible disruption of normal cell morphology and employ a range of drug concentrations in the *in vitro* susceptibility assay that is not cytotoxic. Cytoxicity in cells treated with drug can also be measured by performing viable cell counts using a 0.4% solution of trypan blue and a hemacytometer. The drug concentration required to reduce the viable cell count by 50% is determined by comparison with cell counts in control cultures without drug.

8. For each assay performed, determine the nonspecific background counts for the gamma counter employed by counting two empty counting tubes for 2 min.

Results

Calculations

1. Determine the mean reactive value of the nonspecific background for the gamma counter. Subtract this value from the values of each control and specimen to obtain the net values.
2. Determine the mean reactive value of the three negative control wicks.
3. Determine the mean reactive value of each concentration of antiviral agent.
4. Subtract the mean value of the negative controls from the mean value of each concentration of antiviral agent to establish a net reactivity for each drug concentration examined.
5. Then generate a dose—response curve from which the ID_{50} value for the antiviral agent is calculated using a suitable linear regression probability program.

Interpretation

The ID_{50} value is defined. as the concentration of antiviral drug required to reduce the HSV-specific DNA by 50% compared with the untreated virus controls. In the Clinical Virology Laboratory at Children's Hospital of Philadelphia, HSV isolates with *in vitro* ID_{50} values of ≥2 μg/ml for acyclovir and ≥100 μg/ml for foscarnet are considered resistant to these antiviral agents. Typically, sensitive HSV-

1 isolates have mean ID_{50} values of 0.12 μg/ ml and 35.0 μg/ml for acyclovir and foscarnet, respectively, whereas sensitive HSV-2 isolates have mean ID_{50} values of 0.5 μg/ml and 30.0 μg/ml. The range of ID_{50} values for acyclovir-resistant isolates is from 6.0 to 60.0 *μg/ml.* To date, *in vivo* HSV resistance to foscarnet is rare, but this may become a clinically relevant problem as the use of foscarnet against acyclovir-resistant HSV infection increases.

Quality Control

The gammma counter used to count the hybridization wicks should be monitored with each assay using a ^{129}I reference source of a known counting rate. A detection efficiency of 75% should be observed for the counter.

The negative and positive control wicks supplied by the manufacturer with each hybridization kit are used to verify the performance of the assay. The negative controls are used to monitor the wash conditions whereas the positive control wicks are used to monitor the hybridization conditions.

1. Typical net negative control values range from 250 to 450 counts per minute (cpm). Two of the three negative controls must have net values that are not in excess of 750 cpm to validate the efficiency of the wash.
2. Typical net positive control values range from 10,000 to 20,000 cpm. Both positive controls must have net values above 7,500 cpm to validate the wash incubation temperature. A substantial decrease in the control values will be observed if the wash temperature exceeds 73°C.

Typical net values for the "no drug" virus controls range from 20,000 to 30,000 cpm.

Acyclovir- and foscarnet-sensitive and -resistant control strains of HSV should be tested with each susceptibility assay. The ID_{50} values for each control should fall within an expected range for the assay to be valid. Alternatively, the ID_{50} values for each control should not exceed two standard deviations from mean values established in previous tests, or the assay should be repeated. Table 4 describes the strains and expected ID_{50} values employed for the hybridization assay. The drug-resistant strains of HSV were originally plaque purified to achieve homogeneous populations of resistant viruses. The resistant strains are generally quite stable and do not routinely revert with subpassage. However, it is recommended that continuous subpassage of the control strains be avoided. These isolates should be frozen at –70°C in single-use aliquots and be regrown only when needed.

Limitations of the Procedure

Numerous methodological factors can influence the results of any antiviral susceptibility assay including the DNA-DNA hybridization procedure. These include:

1. The strain of virus examined ("wild" versus "laboratory adapted"). Continuous subpassage of clinical isolates of HSV should be avoided prior to performing susceptibility testing so the isolate tested is representative of the wild-type virus initially identified in a clinical specimen.
2. Heterogeneity of the virus population. A given isolate of HSV may be a heterogeneous population containing both susceptible and resistant viruses, which may make *in vitro* results difficult to interpret. Also, different body sites from a patient may yield different strains of the same HSV type with dramatically different susceptibility patterns.
3. Size of the virus inoculum.
4. Type of host-cell system employed, the age and confluency of the cells when used in the assay, and the cell passage level.
5. Range of drug concentrations employed.
6. Incubation time for virus adsorption and length of incubation of infected cells with drug prior to determination of the ID_{50} value.
7. The use of ID_{50} or $ID_{\%}$ drug end points. A consensus has not been reached concerning which value is most appropriate.
8. Lack of assay standardization between laboratories, making it difficult to compare results. All assays require the ability of the virus to replicate *in vitro* in a host cell system, but each laboratory may use different parameters of viral replication (e.g., CPE, plaque formation, virus yield, viral antigens, viral nucleic acid, viral enzyme activity, or cell transformation) to measure the activity of an antiviral agent. Comparisons of *in vitro* susceptibility data among different laboratories are valid only if the same assay, virus inoculum, and cell type are used. The definition of a sensitive or resistant isolate may differ for each assay system and for different laboratories performin the same assay. When comparing the activity of different antiviral agents against a virus isolate, the same assay system should also be used.

9. Absence of well-characterized, drug-resistant, and -sensitive control strains. These strains are not widely available, but may be obtained as gifts from fellow investigators. Currently, they cannot be purchased through commercial sources.

Certain clinical information should be considered when interpreting the results of *in vitro* antiviral susceptibility assays.

1. Antiviral susceptibility testing appears to be indicated for patients in whom resistance tends to appear the most. These patients normally include immunocompromised individuals, especially transplant recipients and AIDS patients previously treated with anti-HSV agents. One report suggests recurrent acyclovir-resistant HSV infection in an immunocompetent individual which correlated with clinical failure of antiviral therapy.
2. Clinical situations that favor the development of resistance include long-term suppressive therapy, recurrent intermittent therapy, and the use of less than optimum doses of the antiviral agent.
3. Resistance of HSV isolates to acyclovir is becoming more commonplace, occuring at rates of 2-14% in the immunocompromised host. In these patients, resistant HSV can be associated with severe progressive disease that does not respond to antiviral treatement.
4. Data suggest that *in vitro* resistance may predict treatment failure and lead to the institution of alternative antiviral therapy.
5. Antiviral susceptibility testing may become useful in choosing between continuous high-dose acyclovir and foscarnet to treat acyclovir-resistant HSV. As more antiviral agents are developed and utilized, *in vitro* results may also guide the physician in the choice of the most appropriate therapy.
6. Results of *in vitro* susceptibility testing are not always predictive of a clinical response. HSV isolates resistant to acyclovir *in vitro* have responded to antiviral treatment and patients with isolates that remain susceptible to the drug *in vitro* have failed therapy.
7. Acyclovir-resistant clinical isolates of HSV have also been obtained from individuals before therapy and from healthy individuals on chronic suppressive therapy, although no correlation was established between these isolates and the clinical outcome.
8. Acyclovir-resistant isolates may become latent and cause recurrent disease. In some patients, recurrences are caused by drug-sensitive isolates after the successful treatment of infection with a resistant virus.

9. The absolute ID_{50} value for a single isolate of HSV may be less important than the relative change in ID_{50} values between paired isolates.
10. In general, HSV-2 strains exhibit higher ID_{50} values for acyclovir than HSV-1 strains.

As the use and number of effective antiviral drugs increases, a definite need will arise for antiviral susceptibility testing, particularly when patients fail to respond to appropriate therapy. The described DNA-DNA hybridization assay is rapid and reproducible, and provides a convenient and objective method for determining the susceptibility of virus isolates to antiviral agents. Although an ^{125}I-labeled probe is currently employed in this procedure, effective nonradiometric probes are being developed and will offer the advantages of being more versatile, being less hazardous to personnel, having a longer shelf-life, and requiring simpler waste disposal. The commercial availability of the hybridization assay and its relative ease of performance should facilitate standardization and the increased use of antiviral susceptibility testing in diagnostic laboratories.

11

BACTERIOPHAGES

Bacterial viruses, known as bacteriophages or phages (from the Greek *phagein*, 'to eat'), were discovered independently by Frederick Twort (1915) and Felix d'Herelle (1917). A diversity of phages has subsequently been identified and grouped into a number of families. Phage diversity is reflected in both morphological and genetic characteristics. The genome may be DNA or RNA, single- or double-stranded, circular or linear, and is generally present as a single copy. Morphology varies from simple, icosahedral and filamentous phages to more complex tailed phages with an icosahedral head. The majority of phages are tailed.

Phages are common in most environments where bacteria are found and are important in regulating their abundance and distribution. The host controlled modification and restriction systems of bacteria are presumed to protect against phage infection: restriction is levelled against invading double-stranded phage DNA, whilst self-DNA is protected by the modification system. However, in response, certain phages have evolved anti-restriction mechanisms to avoid degradation of their DNA by restriction systems.

Broadly, phages can be classified as either virulent or temperate. A virulent phage subverts the cellular apparatus of its bacterial host for multiplication, typically culminating in cell lysis (for obligately lytic phages) and release of progeny virions. In rare cases, for example the filamentous ssDNA phage M13, progeny are continuously extruded from the host without cell lysis. Accordingly, M13 has been referred to as a 'chronically infecting' phage. Temperate phages have alternative replication cycles: a productive, lytic infection or a reductive infection, in which the phage remains latent in the host, establishing lysogeny.

The latter generally occurs when environmental conditions are poor, allowing survival as a prophage in the host (which is referred to as a lysogen). During lysogeny the phage genome is repressed for lytic functions and often integrates into the bacterial chromosome, as is the case for phage lambda (λ), but it can exist extrachromosomally, for example phage P1. The prophage replicates along with the host and remains dormant until induction of the lytic cycle. This occurs under conditions that result in damage to the host DNA. The phage repressor is inactivated and the lytic process ensues. Such a mechanism allows propagation of phages when host survival is compromised. A resident prophage can protect the host from superinfection by the same or similar strains of phages by repressing the incoming phage genome (a phenomenon known as *superinfection immunity*).

Some temperate phages contribute 'lysogenic conversion genes', for example diphtheria or cholera toxin genes, when they establish lysogeny, thereby converting the host to virulence. Phages can also mediate bacterial genome rearrangements and transfer non-viral genes horizontally by transduction. Tailed phages are the most efficient particles for horizontal (lateral) gene transfer, with the tail effectively guiding injection of DNA into the bacterial cell. Such phage activities generate variability and are a driving force for bacterial evolution.

Bacteriophages have had key roles in developments in molecular biology and biotechnology. They have been used as model systems for animal and plant viruses, and have provided tools for understanding aspects of DNA replication and recombination, transcription, translation, gene regulation and so on.

The first genomes to be sequenced were those of phages. Restriction enzymes were discovered following studies on phage infection of different hosts and laid the basis for the development of gene cloning. Phage-encoded enzymes and other products are exploited in molecular biology. Certain phages have been adapted for use as cloning and sequencing vectors and for phage display. Phages are utilized in the typing of bacteria, in diagnostic systems, as biological tracers, as pollution indicators and in food and hospital sanitation.

There is also renewed interest in the therapeutic potential of phages, due largely to the rapid emergence of antibiotic-resistant bacteria and of new infectious diseases. However, realization of this potential will depend on a number of factors including improved methods of large-scale production and purification of phages, appropriate protocols for administering phages and modifications to enhance therapeutic properties, to remove phage-encoded toxins, to avoid clearance of phages by the host defence system and so on.

RNA Phages

Single-stranded RNA phages

Single-stranded RNA phages are small, icosahedral viruses of the family *Leviviridae* (from the Latin *levis* 'light'), discovered by Tim Loeb and Norton Zinder in 1961. Phages in this family have high mutation rates and some of the smallest RNA genomes known. They are plus strand viruses (with the genome acting as mRNA), containing only a few genes, and infect various Gram-negative bacteria, including *E. coli*, *Pseudomonas* spp. and *Caulobacter* spp. Those infecting the enterobacteria do so by way of the sex pilus. Representative ssRNA phages of the genus *Levivirus* serogroup I (e.g. MS2 and f2) and *Allolevivirus* serogroup III (e.g. Qbeta, Qβ) are considered here.

Virion structure of ssRNA phages

RNA phages typically comprise 180 molecules of major coat (capsid) protein (CP) per virion, one molecule of maturation (A) protein, required for infectivity and maturation, and a linear ssRNA genome of about 3500–4200 nucleotides that displays considerable secondary structure.

Genome of ssRNA phages

MS2 was the first virus identified that carried its genome as RNA. The genome is 3569 nucleotides, with four open reading frames (ORFs) for major coat protein, maturation (A) protein, replicase (subunit II) and lysis protein, and is characterized by intergenic space.

Lysis protein is encoded by an overlapping ORF between the distal part of the coat protein gene, which appears to be required to regulate expression of the lysis gene, and the proximal part of the replicase gene, which encodes the essential residues for functioning of the lysis protein. Occasionally, ribosomes traversing the coat protein gene fail to maintain the correct reading frame so that once translation of coat protein terminates they can reinitiate at the start of the lysis protein ORF, by shuffling a short distance. Efficiency of translation of the lysis protein ORF by such reinitiation is low. Thus only small amounts of lysis protein are synthesized, in turn guaranteeing that sufficient copies of coat protein will be available for assembly of virus particles before cell lysis occurs.

The genome of Qβ, which is about 4000 nucleotides, is larger than that of MS2. It encodes four proteins: coat protein, replicase, maturation (A2) protein, which also mediates cell lysis, and a minor coat protein (read-through protein, A1), which is generated as a result of reading through the coat protein gene into the intergenic region. Read-through is due to a leaky coat protein terminator (UGA). This allows a low level of misincorporation of tryptophan at the terminator signal and subsequent ribosome read-through. The read- through protein constitutes three to seven per cent of the virion protein and has a role in host infection. Such inefficient termination of translation is not found in the group I phages, where there are tandem stop codons (UAA, UAG) at the end of the coat protein gene and no read-through protein is produced.

Not only does the ssRNA genome encode the phage proteins, it also determines specific secondary and tertiary structure that regulates translation, replication and other functions.

Replication cycle of ssRNA phages

For coliphage MS2, infection involves attachment to pilin (the pilus subunit) along the length of a sex (F) pilus of a susceptible host via the A protein. Such binding of phages can block conjugation in *E. coli*.

The plus strand RNA genome directs protein synthesis immediately upon infection. It adopts a complex secondary structure and this largely controls gene expression via differential access of the ribosome binding sites (RBSs) to host ribosomes during translation. Initially the RBS for the coat protein gene is accessible, occupying the loop position of a hairpin structure, whilst those for the maturation (*A*) and the replicase gene are not, being embedded in secondary structure. The replicase gene is expressed only after the coat protein gene has been translated. Such translational coupling between coat protein and replicase genes involves the Min Jou (MJ) interaction. This is a long distance interaction (LDI), in which a sequence just upstream of the replicase gene RBS is base-paired to an internal sequence of the coat protein gene. The MJ interaction represses translation of the replicase gene. However, passage of ribosomes beyond the first half of the coat protein gene temporarily melts this interaction, opening up the RBS for translation of the replicase gene. The rate of re-folding of MJ in part determines the level of expression of replicase. Another LDI, the van Duin (VD) interaction, which borders the MJ, also appears to have a role in repressing replicase translation.

After about 10 to 20 minutes, translation of the replicase gene stops, due to repression by coat protein, ensuring that only a catalytic amount of replicase is translated. Too much replicase could poison the host. The latter half of the infection cycle is then devoted to synthesis of coat protein, which is required in large quantities for virion assembly. A small amount of lysis protein is synthesized during translation of the coat protein gene, due to an occasional reading frame error.

Once replicase is synthesized, replication of the genome can occur; the genome thus switches from a template for translation to a template for replication. There would be a topological problem if replication and translation occurred simultaneously, since they proceed in opposite directions along the template (3' → 5' and 5' → 3' respectively). However, it appears that translation of coat protein is repressed by replicase, which may block the coat protein ribosome entry site, and then recommences when sufficient (-) RNA has been synthesized. The replicase associates with a number of host proteins to form a phage RNA-specific polymerase, which makes both (+) and (-) strands of RNA. Replication occurs in two stages and involves replicative intermediates (RIs) composed of multiple, nascent RNA strands synthesized from a single template, and replicase: RI-1 with (+) RNA as template, and RI-2 with (-) RNA as template. Thus minus strands act as templates for synthesis of plus strands for use as mRNA. Secondary structure, which forms as each new strand is synthesized from the template strand, serves to keep the two strands apart, in turn preventing formation of a double-stranded intermediate. It is noteworthy that the error rate for such RNA replication is quite high compared with that for DNA replication. Replicase is error prone and the lack of a double-stranded intermediate means there is no template for error correction.

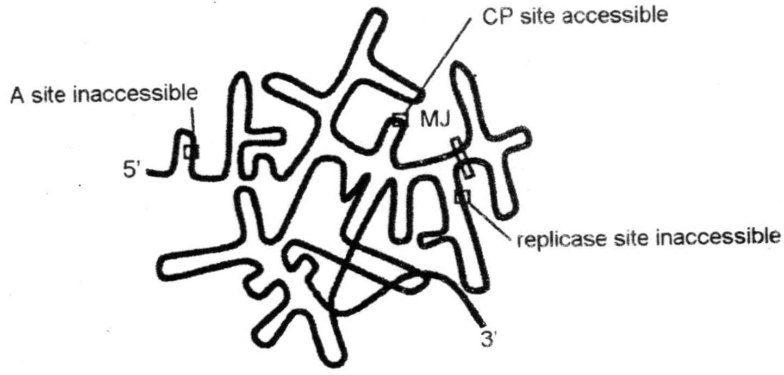

Fig. 11.1. Gene expression in ssRNA phage.

Expression of the *A* gene occurs independently of the others and is limited to periods of nascent plus strand synthesis. The RBS of the *A* gene is masked by folding of the RNA genome through an LDI in which the Shine-Dalgarno sequence of *A* pairs with a complementary sequence upstream in the 5' untranslated leader region. The initiator site of *A* becomes exposed transiently when plus strand synthesis begins, due to a delay in folding of the newly synthesized strand. This allows ribosomes to translate the *A* gene until the RBS is sequestered. The number of molecules of A protein is thus maintained in line with the number of new RNA plus strands. There appears to be up-regulation of the A protein in Qβ, consistent with its role in cell lysis as well as infectivity. The LDI is further downstream than in MS2, so that the RBS is exposed for longer, resulting in higher levels of *A* gene expression.

Phage assembly involves spontaneous aggregation, in which the (+) RNA associates with the A protein and is encapsidated through specific recognition of the phage RNA by coat protein dimers. Progeny virions are normally released by cell lysis, with a burst size of about 10^4, compared with a few hundred for DNA phages.

Thus the minimalist genome of the ssRNA phages is utilized very efficiently. Coding capacity is expanded through the use of overlapping genes. Host proteins are utilized to activate phage proteins and some phage proteins have multiple functions. The phage genome serves as a model system for investigating such mechanisms as translational coupling, translational repression and the role of secondary structure of mRNA in the control of gene expression.

Double-stranded RNA Phages

Phages in the family *Cystoviridae* (from the Greek *kystis* 'bladder, sack') contain a dsRNA genome, which is segmented and packaged in a polyhedral inner core with a lipid-containing envelope. Phi6 (φ6) was the first member of the family to be isolated and has been extensively studied. The genome comprises three linear segments: RNA L (large) of about 6400 nucleotides, RNA M (medium) of about 4000 nucleotides and RNA S (small) of about 3000 nucleotides.

Transcription of the double-stranded genome (for synthesis of new plus strands) involves a phage RNA- dependent RNA polymerase. The plus strand transcripts serve as replication templates and mRNAs. Translation of the L segment produces the early proteins that assemble to form the polymerase complex; the M segment produces structural proteins for membranes and spikes in particular, and the S segment produces structural proteins for the capsid, membrane assembly, lysis and entry, and a non-structural protein for envelopment of the capsid.

Phage φ6 infects its host, *Pseudomonas syringae* (pv. phaseolicola), by way of the pilus, which retracts to bring the virion into contact with the cell, and the nucleocapsid enters. Uncoating occurs inside the cell and the polymerase is released. Transcription of the genome is temporally controlled and virions assemble in the cytoplasm, with their envelope deriving from the host. Packaging of the plus strands occurs in the order S–M–L. These single-stranded precursors are then replicated into mature double-stranded genomes inside the capsid. About 100 virions are released following cell lysis.

DNA Phages

Single-stranded DNA Phages

There are two groups of ssDNA phages, icosahedral and filamentous. Representative icosahedral phages, φX174 and S-13, were amongst the first studied. Later in the 1960s filamentous phages were isolated by Don Marvin and Hartmut Hoffmann-Berling and by Norton Zinder and co-workers.

Icosahedral ssDNA phages

Icosahedral ssDNA phages belong to the family *Microviridae* (from the Greek *micros*, 'small'). Such phages provided the first evidence for overlapping genes and revealed the economy of genetic coding, which is a feature of several small viruses, including hepatitis B virus and the ssRNA phages. Studies on replication of these phages led to the discovery of rolling circle replication and to the identity of various genes encoding proteins for host DNA replication. φX174 has been most extensively studied.

Virion structure of phage φX174

The virion of φX174 contains proteins F, G, H and J; F forms the main shell and the spike protein G protrudes on the icosahedral five-fold axes. A molecule of H is found on each spike and acts as a pilot protein. Part of H may lie outside the shell and part within, spanning the capsid through the channels formed by the G proteins. The projections are involved in attachment to the host and delivery of the genome into the host cell. Highly positively charged DNA binding protein J is associated with the ssDNA genome.

Genome of phage φX174

The genome is a circular ssDNA molecule of 5386 nucleotides, coding for 11 proteins. It was the first entire DNA genome to be sequenced. The genes are tightly clustered with little non-coding sequence. The length of the genome is smaller than its coding capacity. However, the different proteins are generated through extensive use of overlapping genes translated in alternative reading frames or employing different start codons. Gene *A* contains an internal translation initiation site to encode protein A*, which corresponds to the C-terminal region of protein A; *B* is encoded within *A* in a different reading frame; *K* is at the end of gene *A* and extends into gene *C* and is translated in a different frame to *A* and *C*; *E* is totally within *D*, but in a different reading frame; the termination codon for *D* overlaps initiation codon for *J*. All genes are transcribed in the same direction.

Replication cycle of phage φX174

Phage φX174 recognizes the receptor lipopolysaccharide in the outer membrane of rough strains of *Enterobacteriaceae*, such as *E. coli* and *Salmonella typhimurium*, by way of protein H. This protein then acts as a pilot involved with delivery of the phage DNA into the host cell for replication.

Like M13, replication of the genome of φX174 occurs in three stages and involves a double-stranded intermediate, the replicative form (RF). The viral plus strand is first converted to the RF by host enzymes. However, this process is more complex than for M13, involving the formation of a primosome, that includes a number of proteins to open a hairpin found at the origin (*ori*) of replication of φX174 and to load the replication apparatus on to the DNA. The minus strand is transcribed for synthesis of the phage-encoded proteins. Replication of the RF involves rolling circle replication and requires phage-encoded protein A to synthesize new plus strands. These then serve as templates for minus strand synthesis to generate the new RFs. The third stage is asymmetric replication of progeny ssDNA plus strands. RF synthesis continues until sufficient structural proteins have been synthesized and assembled into empty precursor particles. This is mediated by scaffolding proteins (D and B), which associate transiently with the coat proteins and induce conformational switches to drive the assembly process. A procapsid forms, with D forming the external shell and B occupying an internal location. Viral DNA plus strands are then captured, before they can be used as templates for further minus strands, and are packaged with basic protein J (which neutralizes the negatively charged genome). J protein binds both to the DNA and to the internal surface of the capsid, displacing the B protein, to generate the provirion. Subsequent shedding of the D protein yields the mature infectious virion.

Virions accumulate in the cytoplasm and are released by cell lysis, which requires expression of the *E* gene for endolysin production.

Filamentous single-stranded DNA phages

Filamentous ssDNA phages are in the family *Inoviridae*. The F-specific filamentous (Ff) phages, notably M13, fd and f1, have been most extensively studied. They are plus strand phages and are 'male-specific', infecting *E. coli* strains containing the conjugative plasmid F, by adsorbing to the tip of the F pilus (encoded by the plasmid). Unlike many other DNA phages, filamentous phages do not inject their DNA into the host cell; rather entire phage particles are ingested. Furthermore, these phages do not lyse infected cells, but progeny are continuously extruded through the cell membrane. Despite this, plaque-like zones are formed in a lawn of sensitive bacteria, since infected cells grow more slowly than uninfected cells.

Genome of Ff phages

The genome of the Ff phages is a circular ssDNA molecule of about 6400 nucleotides. M13, fd and f1 are 98 per cent identical and generally no distinction will be made between them here. The genes are tightly packed on the genome, with a number of intergenic (IG) regions. The major IG region of 507 nucleotides (between genes *IV* and *II*) contains a number of regulatory regions: the origin of replication for (+) and (-) DNA synthesis, a strong rho-dependent transcription terminator, after gene *IV*, and a 78-nucleotide hairpin region, which is the packaging signal (PS) for efficient phage assembly. There is also a strong rho-independent terminator, after gene *VIII*, and a weaker rho-dependent terminator within gene *I*.

The genome encodes 11 proteins. Genes are grouped by function: genes *II*, *V* and *X* for phage DNA synthesis; genes *III*, *VI*, *VII*, *VIII* and *IX* for capsid structure and genes *I*, *IV* and *XI* for assembly.

Transcription is complex, involving several strong promoters. The two strong transcription terminators divide the genome into two transcription regions: a frequently transcribed region from gene *II* to *VIII* and an infrequently transcribed region from gene *III* to *IV*. All the genes are transcribed in the same direction and there is a gradient of transcription with genes near to a terminator being expressed more frequently, e.g. genes *VIII* and *V*. The weak terminator in gene *I* down-regulates expression of the gene, ensuring that large amounts of gene *I* protein (pI), which would be deleterious to the host, are not synthesized. Gene *XI* protein (pXI) results from an in-frame translation initiator site at residue 241 of gene *I*, and corresponds to the C terminus of pI. Likewise gene *X* protein (pX) is encoded within gene *II*. Such mechanisms ensure that the quantity of each phage protein synthesized is in line with requirements.

Virion structure of Ff phages

The capsid is a flexible protein filament of variable length, depending on genome size; the wild-type length is approximately 900 nm. The filament comprises about 2700 copies of the 50 amino acid major coat protein (gene *VIII* protein, pVIII), arranged in an overlapping array to form a helix. The basic positively charged C-terminal domain of pVIII forms the inside wall of the filament and interacts with the DNA. The N-terminal negatively charged domain is exposed on the outside, giving a negatively charged surface. Hydrophobic interactions hold adjacent coat protein subunits together. The minor coat proteins are present at the ends of the filament. At one terminus (where assembly terminates) are two of these proteins: gene *III* protein (pIII) and gene *VI* protein (pVI), that together form an adsorption complex for recognition of the sex pilus. At the other terminus (where assembly initiates) are gene *VII* protein (pVII) and gene *IX* protein (pIX). The single-stranded circular supercoiled genome is housed lengthwise in the centre of the filament, orientated so that the hairpin packaging loop is located at the pVII/pIX end.

Replication cycle of Ff phages

The receptor for Ff phages comprises the tip of the F pilus together with TolA, TolQ and TolR membrane proteins of *E. coli*. The phage binds first to the tip of the F pilus through domain N2 of pIII. The pilus retracts bringing pIII to the periplasm and domain N1 of pIII can then interact with TolA.

The phage particle enters the cell and disassembles. TolA, TolQ and TolR proteins mediate de- polymerization of the coat proteins; the major coat protein and possibly other capsid proteins being deposited in the cytoplasmic membrane to be re-cycled for new particle formation, as the phage DNA is translocated into the cytoplasm. pIII appears to enter the cell and have a role in the initiation of replication. The circular ssDNA is coated with an *E. coli* ssDNA binding (ssb) protein to protect against degradation by host nucleases. Phage DNA is plus strand and DNA replication occurs immediately upon infection in order to provide the (–) DNA for transcription.

Replication occurs in three stages. The first stage is (–) DNA synthesis to generate the double-stranded RF; this depends on host functions. The minus strand can then be transcribed and translated into phage proteins. A specific endonuclease (gene *II* protein, pII) synthesized from the first RF is required for the second stage of replication, which involves multiplication of the RF. The plus strand of the RF is nicked in the major intergenic region, providing a primer (3' end) for synthesis of a new plus strand, using the minus strand as template, by rolling circle replication. As DNA polymerase III extends the 3' end, the old plus strand is displaced, cleaved and circularized by pII and converted to a dsDNA (RF), which accumulates as a pool in the cell. As DNA replication proceeds there is concomitant synthesis of phage proteins. Gene *V* protein (pV) accumulates in the cytoplasm and at a critical concentration binds cooperatively to displaced plus strands as they roll off the circle, diverting them from their template role. This initiates stage 3 of replication, in which there is a switch from RF to ssDNA synthesis. The 78- nucleotide hairpin region of the plus strand remains free of pV and initiates the phage assembly reaction. pV additionally serves as a translational repressor of pII, in turn regulating DNA synthesis. Another protein, pX, also located in the cytoplasm, has a role in synthesis of the single-stranded viral DNA. The nucleoprotein complex of pV dimers and ssDNA forms a filamentous structure and is targeted to the cell membrane, where assembly takes place. Capsid proteins are inserted into the cytoplasmic membrane as they form.

Assembly proteins pIV, pI and pXI are integral membrane proteins. pIV forms an extrusion pore in the outer membrane and interacts with pI and pX1 of the inner membrane to form a channel across both membranes, large enough to accommodate extruding phage particles. Assembly occurs where inner and outer membranes appear in close contact at an extrusion site and requires ATP, proton motive force and the bacterial protein, thioredoxin. pV is displaced by coat proteins, which are packaged around the DNA as the phage is extruded through the membrane. The process is initiated by interaction of the packaging signal with pVIII, pVII and pIX, possibly directed by pI. A conformational change induced in pI may allow interaction with pIV to open the gate in the pIV exit pore. The pVII–pIX end of the particle emerges first.

Elongation proceeds as pV is displaced by pVIII, whose positively charged C-terminal region interacts with the negatively charged sugar–phosphate backbone of the DNA. Termination of assembly involves pIII and pVI added to the other end of the filament as it is released, the length of the particle being dictated by the size of the viral DNA. Without pIII, elongation would continue with pVIII encapsidating another viral DNA to generate a polyphage. Thus pIII and pVI are important in maintaining structural integrity and stability of the virion. Particles are continuously extruded without killing the host cell.

This assembly process, occurring in the cell membrane and involving polymerization of coat proteins around the Ff phage genome, is in marked contrast to that for dsDNA phages, such as T4, which occurs in the cytoplasm and uses pre-formed heads for packaging DNA.

The filamentous phage CTXφ, which encodes the cholera toxin (the main virulence factor of *Vibrio cholerae*), and a phage associated with invasive meningococci are related to M13. However, unlike the coliphage, these phages can establish lysogeny by site-specific integration into the host chromosome in a similar manner to that of the lambdoid phages.

Cloning vectors and phage display

The relatively small size and non-lytic infection cycle make the filamentous phages easy to manipulate and use as cloning vectors. Furthermore, since filament size is governed by size of the genomic DNA, insert DNA of variable length can be cloned. Vectors are based on the RF, which is easily obtained from infected cells. Cloned ssDNA can be recovered from the phage particles for use in, for example, DNA sequencing, designing DNA probes and site-directed mutagenesis.

The intergenic (IG) region has been manipulated for the development of cloning vectors. Joachim Messing and co-workers designed a set of vectors, the M13 mp series, by inserting the á-fragment of the *E. coli lacZ* gene containing different polylinkers (multiple cloning sites) into the IG region. Foreign DNA inserted into one of the cloning sites can be identified by insertional inactivation of *lacZ*. Other vectors derived from filamentous phages include phagemids, which combine useful features of plasmids and ssDNA phages. Phagemids enable larger inserts than those used in the M13 mp series to be cloned, double-stranded as plasmids, and amplified and recovered as ssDNA in virions. A phagemid typically contains a selectable marker, an origin of plasmid replication, a cloning region (including polylinker and site for insertional inactivation) and the IG of an ssDNA phage for initiation and termination of DNA replication. Superinfection of the host with an ssDNA helper phage provides phage proteins *in trans* for single-stranded phagemid DNA replication and those for phage capsid and assembly. Defective helper phages can be used to reduce helper phage titre. Various phagemids have been developed, including those for phage display.

In phage display a peptide/protein is fused to one of the phage coat proteins so that it is displayed on the surface of the virion. The DNA sequence encoding the peptide is cloned in the phage/phagemid vector. Display libraries are produced by cloning large numbers of DNA sequences. Displayed peptides of interest can then be selected on the basis of their ability to bind to specific targets by a process of biopanning. This allows high throughput analysis of protein–protein interactions. After several rounds of screening, phages displaying peptides with the tightest binding capacity will be enriched. There is a direct link between the nucleic acid insert (genotype) and the displayed peptide (phenotype).

Early display systems used phage vectors with peptides fused to pVIII or pIII. Only small peptides with six to eight residues could be efficiently displayed; larger peptides compromised phage infectivity. The problem of fusion protein size could be overcome by using phagemid vectors with a single copy of the chosen coat protein gene for fusion, the wild-type copy of that gene being on the helper phage. Both wild-type and fusion coat protein contribute to the phage particles so that the surface is phenotypically mixed, and the phages still infective.

A notable application of phage display technology has been in the engineering of antibodies, in which antibody variable domains have been used to construct combinatorial antibody libraries in phagemid vectors. Affinity binding is employed for rapid selection of antibody fragments binding targets of interest.

Phage display was first developed with filamentous phage M13. However, alternative phage systems, such as those of T phages, lambda (λ) and ssRNA phages, have also been employed. The technology promises to revolutionize the study of protein-protein interactions.

Double-stranded DNA Phages

There is a large variety of dsDNA phages, of which the T phages and λ have been particularly well characterized. T2, T4 and T6 belong to the family *Myoviridae* (from the Greek *mys*, *myos*, 'muscle', referring to phages with contractile tails), T1 and T5 together with ë are members of the *Siphoviridae* (from the Greek *siphon*, 'tube', referring to phages with long, flexible, non-contractile tails) and T3 and T7 are in the *Podoviridae* (from the Greek *pous*, *podos*, 'foot', referring to phages with short, non-contractile tails). A representative of each family is considered.

Phage T4

T4, a virulent T-even phage, is one of the largest phages and was the first prokaryotic 'organism' providing evidence of gene splicing through the presence of introns in the genome. It has a complex morphology, consisting of an elongated icosahedral head with a number of different types of protein, including two accessory proteins, Hoc (highly antigenic outer capsid protein) and Soc (small outer capsid protein) decorating the external surface, a contractile tail, baseplate, tail fibres and pins.

Binding of Soc stabilizes the capsid and may account for the osmotic shock resistance phenotype. The two accessory proteins have been utilized in display fusions with e.g. HIV capsid proteins, to provide vaccine vectors. In addition, the tail fibres can be exploited in phage display and may be used as protein struts for nanodevices.

T4 tail fibres, pins and baseplate are involved in binding to the lipopolysaccharide receptor of the *E. coli* host. The tail sheath contracts,

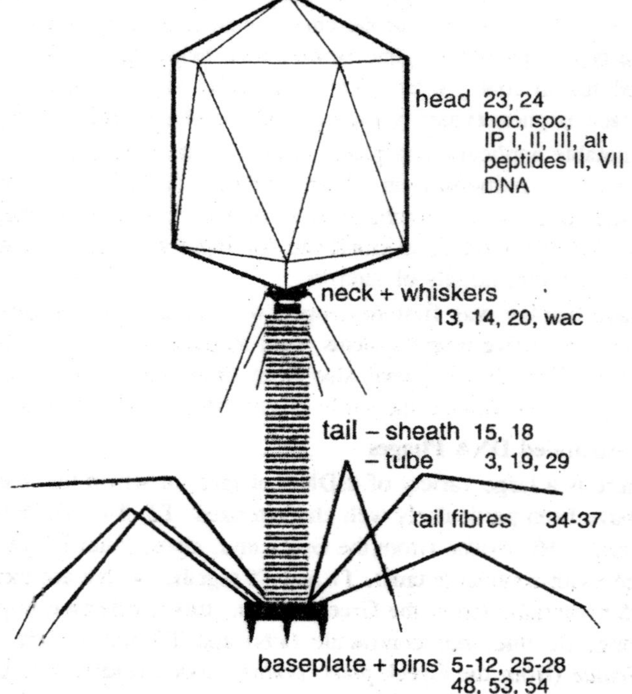

Fig. 11.2. Diagrammatic representation of T4 phage.

driving the internal tail tube into the cell; tail lysozyme (gene product 5, gp5) facilitates digestion of the peptidoglycan layer to reach the inner membrane. The trans-membrane electrochemical potential is required for transfer of the T4 DNA to the cytoplasm. This occurs rapidly, in contrast to the situation with T7 infection. The empty capsid remains extracellular. The phage shuts down the cell; host RNA polymerase stops recognizing host promoters and instead uses phage promoters for T4 gene expression, and the bacterial genome is degraded and recycled in phage DNA synthesis.

The T4 genome is linear dsDNA of about 169 kbp, which is circularly permuted and terminally redundant (a sequence of about 1.6 kbp at one end is directly repeated at the other) due to the mode of replication and packaging by the head-full mechanism. The genome is AT-rich and contains modified bases in the form of 5-hydroxy-methyl-cytosine, rather than cytosine, which protect the phage DNA from many host restriction systems and from phage-encoded nucleases that degrade cytosine-containing host DNA during phage infection. Furthermore, the majority of the hydroxy-methyl groups are glucosylated after DNA synthesis to counteract the host Mcr (modified cytosine restriction) systems, to which the hydroxy-methylcytosine residues confer susceptibility. The genome comprises about 300 probable genes assigned by the timing of their expression, with three different types of promoter: early (Pe), middle (Pm) and late (Pl). The early and middle genes encode functions for DNA replication and for regulating expression of the late genes, which encode head and tail components for the phage particles, and functions for cell lysis.

The regulatory circuit is complex and involves transcriptional termination and anti-termination, transcriptional activators and replication-coupled transcription. Early genes rely on the transcription apparatus of the host, being transcribed from normal σ^{70} promoters. Early gene products alter the host RNA polymerase in two ways for expression of middle genes. The enzyme is thereby able to read through a transcription terminator by an anti-termination mechanism to express genes downstream of early genes and is modified to recognize middle promoters. These promoters differ from σ70 promoters in the –35 sequence, and a transcriptional activator, encoded by the T4 *motA* early gene, allows the host RNA polymerase to transcribe from such promoters.

Transcription of the late genes is coupled to replication. Promoters for these genes again differ from σ^{70} promoters and require an

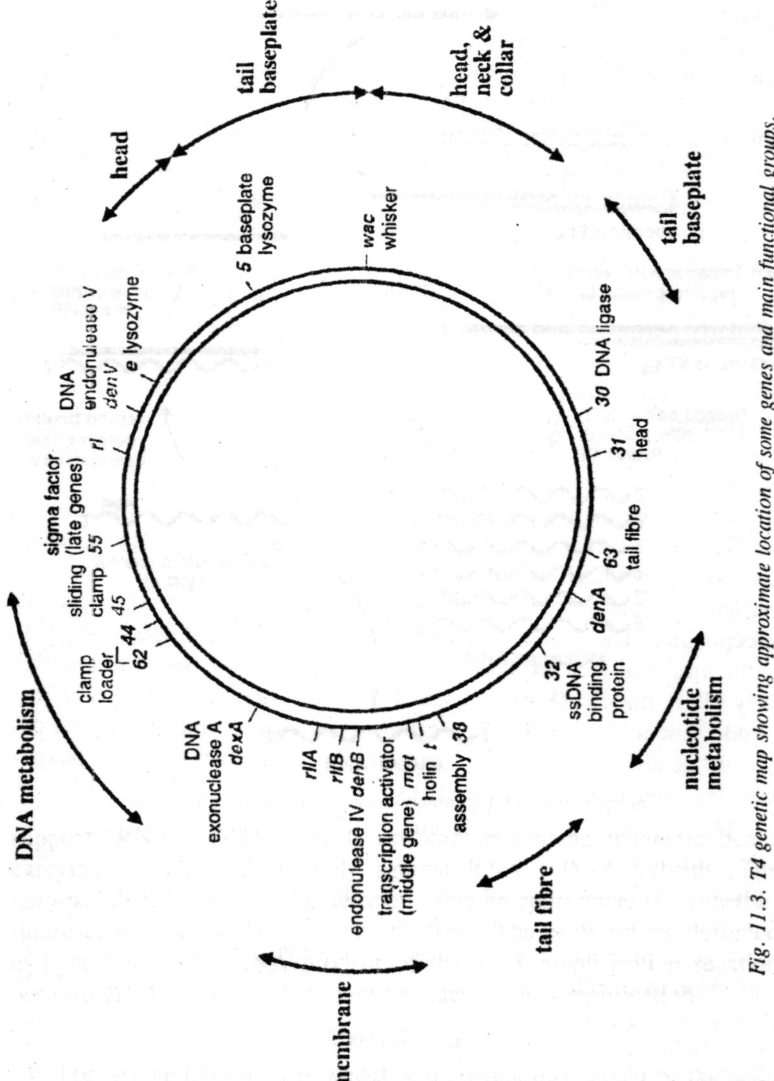

Fig. 11.3. T4 *genetic map showing approximate location of some genes and main functional groups.*

alternative sigma factor, encoded by T4 regulatory gene *55*, to activate transcription. Binding of gene *55* product to host RNA polymerase allows the enzyme to recognize specifically T4 late gene promoters. Other T4 gene products, including those of genes *44*, *45* and *62*, which are involved in replication, are additionally required for late gene transcription. RNA polymerase complexed with gp55 seems to need an actively replicating T4 DNA for effective transcription of the late genes. Such coupling between replication and transcription ensures that phage DNA becomes available for packaging as the capsid proteins are synthesized from the late genes. The DNA is packaged into the pre-made heads. These require a protein scaffold to assemble. Packaging is size specific, depending on the size of the head, and is accompanied by a structural change in the procapsid. Tail and baseplate assemble separately. Heads attach to tails and tail fibres are added last to produce mature progeny phages. T4 holin (gene *t* protein) creates a hole in the inner membrane so that lysozyme (gene *e* protein for endolysin) can degrade the peptidoglycan for cell lysis and release of phages. Upon superinfection of T4-infected bacteria, the phage exhibits the phenomenon of 'lysis inhibition', whereby the latent period is prolonged, lysis is delayed and an increased burst size (about 1000) results. It is noteworthy that T4-like phages have been suggested as possible therapeutic agents in, for example, the treatment of *E. coli* diarrhoea. Lysis of the majority of diarrhoea-causing *E. coli* strains would probably require a cocktail of several phages and/or genetic manipulation to extend host range sufficiently.

Phage T7

T7 is a virulent T-odd phage, with icosahedral head and short, non-contractile tail. The linear genome (about 39 kbp) is smaller than that of T4. It is terminally redundant and comprises about 50 genes, which are closely packed, divided into three groups and numbered from the genetic left end, the first end to enter the host cell. Integral numbered genes are essential; most non-integral numbered are non-essential. T7 early genes (up to 1.3) are transcribed by the bacterial RNA polymerase, which is afterwards inactivated. One of these early gene products is T7 RNA polymerase, a very active enzyme that takes over transcription of T7 genes. Late genes are transcribed from strong T7 promoters. T7 transcription is regulated both by the slow delivery of the DNA into the host and by the synthesis of the T7 RNA polymerase, which recognizes only T7 promoters. This specificity of T7 RNA polymerase has been exploited in the development of a number of expression vectors that incorporate these promoters. Degradation of the bacterial chromosome during infection provides the nucleotides for T7 replication, which uses T7 DNA polymerase. This polymerase has been modified and marketed commercially as the DNA sequencing enzyme, Sequenase.

Phage lambda (λ)

Morphologically, phage λ comprises an isometric head containing a dsDNA genome of about 48.5 kbp and a long, flexible tail with central tail tip fibre and side tail fibres. Genes are clustered according to function on the genome, with many encoding head and tail components of the phage, but only two (*O* and *P*) for replication, unlike phages T4 and T7, where there are many phage-encoded replication proteins. The genome in the virion is linear, with 12 b single-stranded complementary 5' ends (cohesive ends) that mediate circularization of the DNA after infection of the host. Circularization of the λ genome protects it from degradation. The cohesive ends are generated due to the mode of packaging of the DNA from concatemers, (containing multiple λ genomes), which are cut asymmetrically at ë *cos* sites. The specificity of this packaging mechanism has been exploited for efficient introduction of insert DNA into recombinant hosts using λ cloning vectors. To subvert bacterial restriction–modification activities upon infection, phage λ encodes the Ral (restriction alleviation) protein. This enhances methylation of the viral DNA, thereby alleviating restriction by type I restriction enzymes.

As a temperate phage, λ can either establish lysogeny, being maintained as a prophage integrated into the bacterial chromosome between the biotin and galactose operons, or multiply lytically to produce progeny virions upon infection. The decision between the lytic cycle and lysogeny depends on a number of genetic, environmental and physiological factors. Gene expression is negatively regulated by repressor–operator interactions, with termination and anti-termination of transcription playing important roles in temporal control.

Phage λ infection involves the *E. coli lamB* protein receptor. Upon introduction into the host the phage genome circularizes and the bacterial RNA polymerase binds to leftward and rightward promoters (P_L and P_R), transcribing *cro* and *N* genes. N protein, together with host proteins, mediates anti-termination at transcription terminators t_{L1}, t_{R1} and t_{R2}, so that delayed early genes including *cII* and *cIII* are expressed. Anti-terminator protein, N, acts by modifying the RNA polymerase, as it passes through the *nut* (N utilization) sequences (*nutL* and *nutR*), so that it ignores termination. N may additionally impede release of the transcript by the terminator hairpin or by the activity of rho helicase. Where the concentration of CII (transcription activator protein) remains below a critical level, which is the case in the majority of cells, transcription from PL and PR continues until Q anti-terminator switches on late genes governing synthesis of phage components for heads and tails, and *R*, *Rz* and *S* for lysis. The *S* gene protein allows transport of λ endolysin, encoded by *R* and *Rz*, to the periplasm. The *R* gene product degrades the peptidoglycan of the cell wall, with the *Rz* product having a minor role in cell lysis. The effectiveness and specificity of phage lytic enzymes make them promising candidates for use as antibacterial agents ('enzybiotics') in the control of pathogenic bacteria, including antibiotic-resistant bacteria.

Under conditions where CII protein prevails, it activates transcription of *cI* from P_{RE} (promoter for repressor establishment) and of *int* from P_I (promoter for integrase). This results in integration of the λ genome into the bacterial chromosome and repression of lytic functions, through CI repressor binding to O_L and O_R, for establishment of lysogeny. CII also activates P_{aQ} for transcription of antisense RNA that reduces expression of *Q*. Lysogeny is maintained by continued synthesis of CI repressor from P_{RM} (promoter for repressor maintenance). Bacterial proteins, such as HflA protease, and phage-encoded CIII protein affect stability and hence concentration of CII. These proteins are themselves affected by environmental stimuli, e.g. temperature and nutrients.

Physiological stress, effected by e.g. exposure of a λ lysogen to UV irradiation, results in damage to DNA and induces the SOS response, in turn activating RecA co-protease. This stimulates a latent protease activity of CI with subsequent cleavage of the repressor, so that it no longer binds to O_R and O_L. Anti-repressor Cro in turn binds to O_R and O_L to prevent transcription of *cI* and stimulate its own transcription. The phage is thereby induced to enter the lytic cycle. Genes *xis* (for excisionase) and *int* (for integrase) are expressed and the λ genome excises to direct the lytic cycle.

The λ lysis–lysogeny decision represents a paradigm for control of gene expression in bacteria and lysogenic phages. Relatives of phage λ may carry verocytotoxin (VT) genes to convert their hosts into pathogens upon lysogenization, as is the case with *E. coli* O157. Phage λ, which itself carries *bor*, conferring serum resistance on the lysogen, provides a useful model system for studying such lambdoid phages.

Phages of the lambda family and certain filamentous phages are important vectors for spread of virulence genes. Thus, despite potential lethality to their hosts, such phages may confer selective advantages, through e.g. lysogenic conversion and superinfection immunity, to contribute to bacterial fitness in the environment.

12

Detection of Viral Pathogens

The polymerase chain reaction (PCR) was introduced to the scientific community in 1985 by scientists from Cetus Corporation and has since become an integral part of the molecular biology laboratory. Several improvements in the technology and associated equipment have significantly changed the way PCR is performed today relative to the earlier PCR protocols. Well-characterized user-friendly PCR assays designed for diagnostic applications in clinical laboratories have been described for the detection of a variety of human viral pathogens.

PCR is a relatively simple procedure consisting of repetitive cycles of oligonucleotide primer-directed DNA replication. A selected genetic target is copied in a three-step reaction consisting of (1) denaturing the target DNA at elevated temperature (90-95°C), (2) cooling the reaction to promote annealing of oligonucleotide primers that are complementary to either strand of the target DNA, and (3) extending the bound primers by DNA polymerase action, resulting in a replication of the target sequence. The reaction is repeated in successive cycles of heating and cooling, referred to as thermocycies; each cycle doubles the input target sequence. Because the newly synthesized DNA copies can also serve as templates in subsequent replication cycles, a 2^n accumulation of product copies is possible, where n is the number of thermocycles. For this reason, PCR is often referred to as an exponential amplification technology. In practice, less PCR product than is theoretically predicted is actually synthesized because of lower reaction efficiency after many cycles (>25) of PCR amplification.

Several improvements that have enhanced the analytical sensitivity and specificity of PCR procedures have been introduced since 1985. The discovery of thermostable DNA polymerases such as *Taq* DNA polymerase greatly simplified the PCR procedure by eliminating the need to add fresh enzyme after each cycle of denaturation because of inactivation of the thermolabile DNA polymerase enzyme used in earlier PCR protocols. The use of *Taq* DNA polymerase also resulted in increased specificity of PCR amplification reactions by permitting higher primer annealing and polymerase extension temperatures to be used, decreasing the amount of nonspecific target amplification. Implementation of target-specific PCR protocols with optimized reaction buffers and target-specific thermocycling profiles, and incorporation of innovative primer designs capable of accommodating mismatched bases due to sequence heterogeneity furthered the potential usefulness of PCR-based diagnostic procedures. Development of new equipment such as high sample capacity thermocycling machines (microprocessor-controlled heating–cooling blocks), capable of rapidly and accurately attaining target temperatures, increased assay sample throughput and precision. Better liquid handling equipment, including robotic work stations that eliminate the need for individual pipetting of PCR reaction mixes, has improved the cost effectiveness of routine PCR testing by reducing the hands-on time required to perform PCR assays.

Specificity of PCR Assays

The high analytical sensitivity of PCR-based procedures facilitates the detection of minute amounts of pathogens in biological specimens that are often undetectable using more conventional technology. However, the accumulation of high copy numbers of specific target nucleic acid sequences presented problems with controlling the specificity of PCR amplifications. False positive results due to re-amplification of PCR products carried over from previous amplifications of positive specimens and controls plagued early attempts to implement routine clinical PCR testing procedures. The power and potential pitfalls of routine use of an optimized PCR protocol for the detection of human immunodeficiency virus type 1 (HIV-1) proviral DNA. A 100-μl PCR assay can generate up to 10^9 copies of PCR product (amplicons) from only 10 copies of input HIV-1 DNA. Carryover contamination of subsequent reactions with as little as 10^{-6} μl (aerosol) amplification product can result in virtually all false positive reactions.

Physical separation of pre- and post-PCR areas within the laboratory, dispensing of reagents into single-use aliquots, and use of positive displacement pipettors or aerosol-resistant pipette tips helped reduce the possibility of carryover contamination. However, given the stability of unmodified DNA products and the exceedingly low level of carryover contamination that can result in false positive results, many laboratories still experienced some problems controlling reaction specificity because of low-level PCR product contamination. More recently, enzymatic, photochemical, and chemical methods to modify PCR amplification products so they cannot be re-amplified have been described that significantly reduce the possibility of false positive PCR results due to carryover contamination.

Enzymatic Inactivation Using UNG

Incorporation of dUTP and the enzyme uracil-*N*-gylcosylase (UNG) into PCR mixes is a simple and effective means of carryover contamination control. UNG enzyme was first recognized as a component of a DNA excision and repair system observed among many bacterial species. The function of the enzyme is to remove spontaneously arising uracil residues, the result of deamination of cytosine residues, from the DNA chain prior to incorporation of a new cytosine residue. By substituting dUTP for dTTP in PCR mixes, U is substituted for T by *Taq* polymerase during the primer extension (elongation) step of PCR amplification. The resulting product amplicons will contain deoxyuracil residues in contrast to the naturally occurring deoxythymidine residues present in template DNA derived from

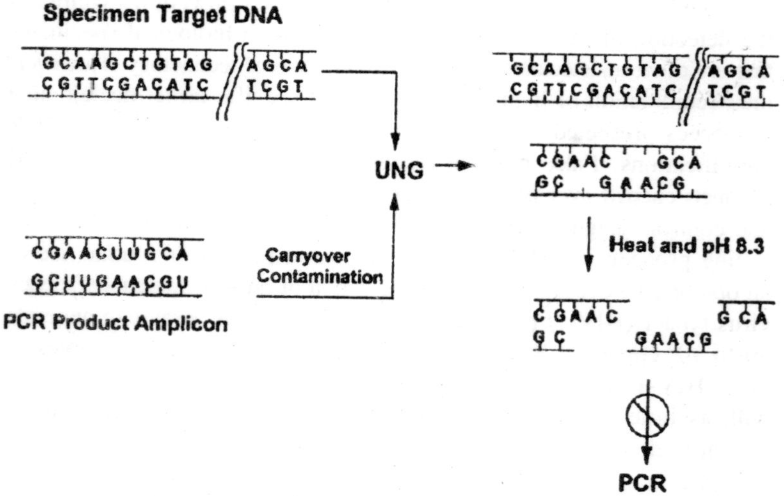

Fig. 12.1. UNG restriction of uracil-containing amplification products.

specimens. As carryover contamination of target nucleic acids (specimens) with deoxyuracil-containing amplicons *prior to* PCR amplification can be controlled by the inclusion of UNG in the amplification reaction mix.

UNG cleaves the uracil residues from the contaminating DNA amplicons, resulting in abasic sites in the carryover DNA. The abasic sites are hydrolyzed under the buffer conditions and elevated temperatures that are optimal for *Taq* polymerase during PCR amplification and thus do not serve as efficient templates for re-amplification. However, newly synthesized deoxyuracil-containing product DNA can still be used for detection by ethidium bromide staining of agarose gels and probe hybridization methods, or as templates for cloning (UNG^- hosts only) and sequencing experiments. Therefore, inclusion of dUTP and UNG enzyme in PCR mixes inactivates any contaminating PCR product DNA and modifies newly synthesized products so they, in turn, can be inactivated rather than serve as templates in subsequent PCR amplifications. When incorporating UNG restriction procedures into PCR protocols, note that UNG denatures at temperatures above 55°C; therefore, it will not cleave newly synthesized products. However, UNG will renature on cooling of PCR assays. To prevent UNG from degrading newly synthesized amplicons and compromising assay sensitivity, reactions must be kept at 55°C prior to product analysis or UNG activity should be destroyed by the addition of NaOH (~ 0.2 *N* final concentration) to establish basic conditions. Inactivation of 10^6-10^7 uracil-containing amplicons is readily achievable using UNG restriction methods. However, some GC-rich products may not be efficiently inactivated. One must optimize PCR mixes that incorporate UNG and empirically determine the efficiency of product inactivation.

The use of UNG in PCR protocols can also improve the sensitivity of PCR amplifications by degrading nonspecific extension products that result from low-stringency primer binding that occurs below the optimal primer annealing temperature. Once the reaction temperature is raised to around 55°C, UNG activity is lost, permitting the accumulation of specific amplified target sequences that are produced following high-stringency primer annealing. In this capacity, and in similar "hot-start" PCR protocols, reaction sensitivity is enhanced because of less production and amplification of nonspecific sequences during the initial PCR cycles. Another benefit of UNG carryover control is that RNA templates can be used with this method because UNG does not cleave uracil residues in ribonucleic acid chains. This feature is particularly advantageous when RNA-PCR procedures are employed for the detection of viral pathogens with RNA genomes or when targeting mRNA for expression studies provided that the reverse transcriptase/ DNA polymerase enzyme(s) employed can use either thymidine or uracil during transcription and amplification.

Using combinations of physical separation of sample preparation and amplification/product detection areas, aerosol-resistant pipettors and tips, and amplification product inactivation methods, successful PCR testing programs have been implemented in both small and large volume clinical laboratories.

Considerations for Diagnostic Assay Design

When applying PCR amplification methods for diagnostic use, one must carefully consider the variables that may influence the potential utility of the assay. The selection of the nucleic acid target for amplification, the design of oligonucleotide primers and probes, the optimization of specimen preparation and PCR amplification conditions, and the method of amplified product detection and inactivation must all be optimized for the intended pathogen. The following discussion provides an overview of areas that must be considered when designing amplification reactions. Readers should keep in mind that each system must be individually optimized.

The selection of an appropriate amplification target for diagnostic applications should begin with a review of the published sequence information available for the organism. Nucleic acid sequence databases such as GenBank are useful in the search for sequence information for isolates of a particular organism or a related group of organisms. Computer disks containing volumes of sequence information and programs to analyze the information are available from several commercial sources. The target selected can be DNA or RNA and should be a genetically stable (conserved) region. When RNA is targeted, a separate reverse transcription step must be included in the assay design to convert RNA to cDNA, which then serves as template DNA for PCR amplification. For diagnostic applications, the target should be between 100 and 600 by in size, have approximately 50% GC content, be devoid of stretches of the same nucleotide, and not possess significant regions of self-complementarity or kinked (hairpin) secondary structures.

Primer Selection

The design and selection of primers are influenced driven by the target sequence. In general, primers should be between 25 and 35 bases in length, avoid contiguous stretches of the same nucleotide, possess melting temperatures (T_m) between 55 and 65°C, and avoid complementarity within a primer and between primers, especially near the 3' ends. Shorter primer sequences (<20 bases) are less capable of accommodating mismatched bases with the target DNA or RNA template and are not recommended for diagnostic applications. A thymine residue can be included at the 3' end of each primer to ensure DNA polymerase extension of a bound primer that is mismatched with the template base at the 3' end of the primer sequence. Therefore, both the overall length and the base composition of primer sequences contribute to the ability to accommodate sequence mismatches and avoid potential false negative amplification results. The size and composition of the resulting PCR product should also be considered during primer selection when amplification product inactivation methods are intended as part of the PCR assay protocol. Selection of primers for PCR amplification protocols has been greatly simplified by the availability of computer programs that accurately predict T_m, secondary structures, and primer—primer interaction.

Detection of Amplification Products

For diagnostic applications, the use of DNA probes for detection of amplified products is recommended over non-probe-based detection strategies such as ethidium bromide staining of DNA following agarose gel electrophoresis or nested amplification using a second set of primers that recognize the initial amplification product. Because specificity of amplification varies widely among amplification protocols, visualization of the expected size DNA fragment following amplification does not guarantee that the specific target amplicon is present. Nested PCR protocols involve the opening and transfer of amplification product tubes, which greatly increases the possibility of carryover contamination. In addition, product inactivation protocols cannot be used in nested PCR systems because the first PCR product must serve as a template for the second (nested) amplification reaction. A single-tube nested PCR method has been described that may circumvent some of the problems associated with diagnostic use of two-tube nested PCR protocols. In general, DNA probe-based detection protocols are rapid and simple, and add another level of specificity to diagnostic PCR protocols for clinical laboratory use. The abundance of the PCR product generated after amplification facilitates detection by a DNA probe by allowing the product concentration, not the amount of labeled probe added to a detection reaction, to drive the detection reaction sensitivity.

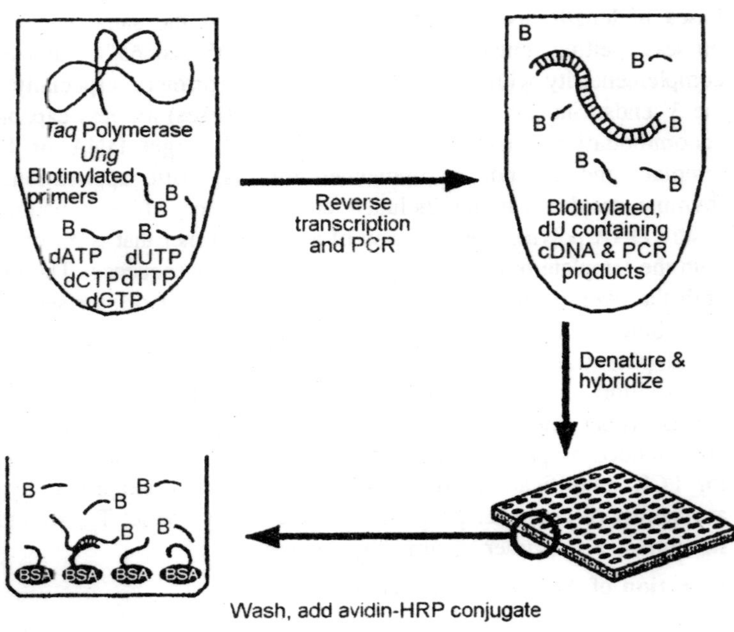

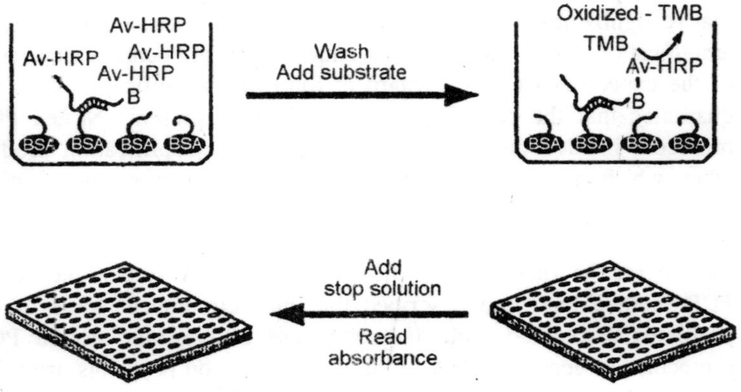

Fig. 12.2. Nonisotopic DNA probe detection format.

The introduction of reporter molecules, or ligands for subsequent detection protocols, into primer or probe sequences has greatly simplified the detection of PCR amplification products. Several rapid and sensitive nonradioactive DNA probe detection protocols have been described for routine clinical laboratory use, including the use of acridinium ester derivatized DNA probes and microtiter plate based probe detection of biotinylated PCR products.

PCR Amplification for the Qualitative Detection of HIV-1 Proviral DNA

The related human retroviruses, human immunodeficiency virus types I and II (HIV-1, HIV-2) are the etiologic agents of acquired immunodeficiency syndrome (AIDS). In North America, HIV-1 is responsible for the majority of AIDS cases whereas infection with HIV-1 or HIV-2 is associated with AIDS in Africa. Infection with HIV-1 and progression to AIDS-related complex (ARC) and AIDS occurs in successive stages of disease. Acute infection is characterized by a short period, usually between 2 and 10 wk, when circulating virus may be detectable (viremia) but no serological evidence of infection (no detectable antibodies) is present.

Recently infected individuals in this "*seronegative window*" period are likely to be highly infectious because of the presence of high titers of circulating viral particles prior to the stimulation of an immune response and the production of anti-viral antibodies (seroconversion). After seroconversion, most circulating viruses are complexed and removed from the circulation. During this stage, which may last for years, most HIV-infected individuals can be reliably detected using standard enzyme immunoassay (EIA) antibody detection methods. Over time, HIV infection leads to a progressive depletion of $CD4^+$ T lymphocytes with associated deterioration of immune system function. Progression to clinical AIDS is characterized by the occurrence of opportunistic infections and other potentially life-threatening illnesses associated with immune system dysfunction.

Direct detection of viral particles, or viral nucleic acid, offers the potential to detect early HIV infection and to resolve inconclusive HIV antibody test results that may occur due to early seroconversion or nonspecific cross-reactivity in the case of uninfected individuals. Direct detection of viral particles or nucleic acid is the most reliable method for diagnosis of HIV infection in newborns of HIV-infected women because of the presence of circulating maternal antibodies in the infant for up to 15 mo.

PCR is ideally suited to these situations because of its ability to amplify HIV-specific nucleic acid sequences from small volumes of blood [e.g., HIV proviral DNA in peripheral blood mononuclear cells (PBMC)1 or plasma (virion-associated HIV RNA). The following protocol describes a method for the amplification and detection of HIV-1 proviral DNA from whole blood specimens.

Protocol: Amplification and Detection of HIV-1 Proviral DNA in Whole Blood Specimens

Materials and reagents

1. Leukoprep blood collection tubes or ACD Vacutainer® tubes
2. Sterile Tris-buffered saline: 10 mM Tris, pH 7.5, 0.15 *M* NaCl
3. Sample extraction buffer: 50 mM KCl, 7.5 mM $MgCl_2$, 0.5% Tween 20, 0.5% NP-40, 10 mM Tris pH 8.3, 100 mg/ml proteinase K
4. 2× PCR master mix: 20 mM Tris, pH 8.5, 100 mM KC1, 400 μM each of dATP, dCTP, dGTP, and dUTP, 25 pmol SK462 (upstream primer), 25 pmol SK431 (downstream primer), 0.05 U/μl *Taq* DNA polymerase (2.5 U/50 μl); store 2× master mix at 4°C; immediately prior to use, add uracil-N-glycosylase (UNG) enzyme to 0.04 U/μl (2.0 U/50 μl).

   Note: 2× PCR master mix is delivered at 50 μl per PCR assay tube. Master mix is diluted to lx by addition of 50 μl sample in sample extraction buffer; final reaction volume is 100 μl.
5. SK462: 5'-AGTTGGAGGACATCAAGCAGCCATGCAAAT-3'
6. SK431: 5 '-TGCTATGTCAGTTCCCCTTGGTTCTCT-3'
7. Positive control DNA: HIV-1 plasmid DNA; dilute to 20 copies/ 50 μl in dH_2O containing 1μg human placental DNA
8. Specimen negative control: HIV negative donor PBMC or "mock" specimen tube (buffers only)
9. SK 102 capture probe: 5'-GAGACCATCAATGAGGAAGCTG-CAGAATGGGAT-3' end-labeled at the 5' end with 32p.

Specimen collection and preparation

Collect 5-7 ml whole blood into a Leukoprep or ACD Vacutainer tube. *Note*: All specimens should be considered capable of transmitting an infectious agent. Universal precautions should be observed when handling specimens. Laboratory coats, gloves, and safety goggles should be worn at all times. Specimens should be processed in a biological containment hood (BL-2 containment).

1. Centrifuge whole blood in leukoprep tubes to separate PBMC. Alternatively, centrifuge whole blood collected into ACD anticoagulant and layered onto Ficoll—Hypaque for 30 min at 400 *g* to separate PBMC.
2. Collect the PBMC (buffy coat), transfer to a sterile plastic 2-ml screw-cap tube, and wash with 1.5 ml Tris-buffered saline, pH

7.5. Centrifuge at 200 *g* for 10 min in a microfuge to pellet the cells.

3. Remove the supernatant. Resuspend the pellet in 1.5 ml Tris-buffered saline and centrifuge at 200 *g* for 10 min in a microfuge.
4. Remove the supernatant and resuspend the pellet in 400 μl sample extraction buffer (final concentration of cells should be set at ~3 × 10^6 cells/ml).
5. Vortex the resuspended pellet and incubate at 60°C for 30 min (water bath or heat block).
6. Transfer to a 95-98°C water bath/heat block for 30 min to inactivate proteinase K.
7. Remove tubes; let cool to room temperature.
8. Store cell lysates in refrigerator.

Note: A specimen preparation negative control should be introduced at Step 3 and processed together with specimens through Step 9.

PCR amplification

Note: This protocol is described for amplifications performed using the GeneAmp 9600 Thermocycler (Perkin-Elmer) with thin wall "*microamp*" tubes (Perkin-Elmer).

1. Dispense 50 μl complete 2× PCR master mix (UNG enzyme added) into each reaction tube according to the number of samples to be amplified, plus a positive (HIV-1 DNA plasmid) and specimen negative control tube. In addition, an amplification control (master mix + 50 μl dH_2O) should be included. All specimens and controls should be amplified in duplicate reactions.
2. Using a positive displacement pipettor or micropipettor and aerosol-resistant tips, add 50 μl specimen lysate to the appropriately labeled reaction tubes (less specimen can be added and the volume made up to 50 μl with glass-distilled water if necessary; target cell number per PCR reaction is ~150,000 cells or ~1 μg genomic DNA). Pipette positive, negative, and amplification controls last.
3. Cap each tube after addition of sample. After each row of 8 tubes is completed, securely cap the row of tubes using the roller tool provided with the GeneAmp thermocycler. Single reaction tubes with individual caps are available and are recommended for small volume uses.
4. Place the tubes in the thermocycler and run the thermocycling profile.

5. Ramp the thermocycler down to 20°C and remove the reaction tray. Denature the UNG enzyme by adding 100 μl 0.4N NAOH, 80 mM EDTA or freeze at –70°C.
6. Detect the amplified HIV-1 DNA using 32P-end labeled SK-102 probe as described by Kwok and Sninsky (1993).

Note: Nonradioactive detection protocols using chemiluminescent and colorimetric methods have also been described and may be preferred by some users.

Analysis of Data and Interpretation of Results

Assay controls

Analysis of the data begins with interpretation of positive and negative controls. Both duplicate amplifications of the 20-copy HIV-1 DNA positive control should yield unambiguously positive signals. Neither the specimen negative control nor the amplification control reaction should produce any positive reaction product. If either or both of the positive control amplifications do not give positive signals, assay sensitivity cannot be guaranteed and the run should be repeated. If either or both of the duplicate amplifications of the specimen negative and amplification controls give signals above a predetermined background level, assay specificity cannot be ensured (potential carryover contamination or specimen mix-up has occurred) and the run should be repeated.

In some instances, repeat analysis of the amplified products can be performed (without re-amplification of the specimen) to rule out pipetting errors during product analysis when negative control values are too high. Assay positive control signal levels must be determined empirically by analysis of known positive and negative specimens, as well as with dilutions of positive control plasmid DNA, to determine the reproducible analytical sensitivity of the system.

Specimen results

When the assay control values are within range, the signals observed for the specimens can be analyzed to determine results. Specimens with clearly positive signals from each of the duplicate amplification reactions are interpreted as positive for HIV-1 proviral DNA. Alternatively, if neither value for a specimen yields a positive signal, the specimen is interpreted as negative for HIV-1 proviral DNA. If discordant results are obtained for the duplicate reactions, the specimen lysate should be re-amplified and detected in duplicate or re-isolated from the original blood specimen tube. If results remain discordant, the result is interpreted as inconclusive and a second specimen should be obtained for repeat analysis.

Several factors can lead to discrepant or false negative results including PCR inhibitors contained within the specimen, sample bias due to low copy numbers of the HIV target within the specimen, and poorly optimized preparation, amplification, or product analysis procedures. A separate control amplification of a resident human gene sequence (e.g., HLA DQ alpha or beta-globin) can be performed to check for the presence of gross inhibitors. Amplification of a prepared specimen in duplicate reactions helps avoid false negative results due to sample bias with low copy number specimens. Depending on volume and cost constraints, single or duplicate amplification reactions from two different preparations of the same specimen further decrease the possibility of not detecting low copy number positive specimens. The specimen preparation and PCR amplification parameters for the assay described here have been optimized and the results obtained from clinical specimens have been described. With proper consideration given to assay design and optimization and implementation of proven PCR quality control procedures, most laboratories should be capable of generating consistent, high-quality results using PCR amplification and DNA probe detection methods.

13

DETECTION OF HUMAN PAPILLOMAVIRUSES

The specificity and quantitative nature of nucleic acid hybridization has been exploited in the past three decades for a variety of purposes. With the advent of nonradioactive detection methods, nucleic acid hybridization has become a major tool in detection of infectious agents or genetic abnormalities. The sensitivity required for diagnosis of many infectious diseases has not been achieved by nucleic acid hybridization assays that do not have an amplification step. Numerous approaches have been developed to overcome this shortcoming, some of which are described in this volume. Enzymatic signal amplification systems have been described that use alkaline phosphatase as the detection system. Other investigators have used nucleic acid probes that carry large numbers of detectable moieties such as alkaline phosphatase, thereby increasing the detectable signal per hybridizable target sequence. Despite some success in the use of these approaches, none has been universally applicable.

An efficient solution to this problem has been the development of *in vitro* methods for amplifying the nucleic acid sequences of interest. The polymerase chain reaction (PCR) was the first method for sequence-specific amplification of target molecules in a sample. This method has rapidly become one of the most important tools in molecular biology. The widespread utility of amplification in molecular biology and molecular diagnostic procedures has stimulated the development of other nucleic acid amplification methods. Although some of these methods are dependent on the presence of a DNA target and amplify a target sequence; other methods are based on replicatable probes that are amplified after hybridization to reveal the presence of the probe. Target amplification systems such as PCR, the ligase-based amplification system (LAR), the ligase chain reaction (LCR), and the transcription- based amplification system (TAS) use repeated temperature cycling to achieve amplification. In constrast, the probe amplification systems and the self-sustained sequence replication reactions do not require thermal cycling. In this chapter we describe an isothermal method for amplification of nucleic acids based on ligation of a promoter sequence to a target nucleic acid, resulting in target amplification through a transcribed RNA intermediate. This target amplification methodology has been combined with solution hybridization-based enzymatic detection of amplified target nucleic acid to produce a highly sensitive method for the detection of nucleic acids. Application of these technologies to detection of human papillomaviruses (HPV) is described.

PRINCIPLE OF LIGATION-ACTIVATED TRANSCRIPTION AMPLIFICATION

The ligation-activated transcription (LAT) reaction requires the simultaneous action of four enzymes in a single isothermal reaction. The enzymes required for this system include DNA ligase, RNA-dependent DNA polymerase (reverse transcriptase), RNA polymerase, and ribonuclease H (RNase H). The LAT procedure is based on the ligation of a double-stranded T7 RNA polymerase promoter sequence to the 3' end of the DNA to be amplified. This novel oligonucleotide, or proto-promoter, contains a blocked 3' single-stranded region that is complementary to the 3' end of the target sequence and a double-stranded T7 RNA polymerase promoter. After ligation of the proto-promoter to the target, the RNA polymerase produces multiple copies of RNA complementary to the DNA. These RNAs are, in turn, converted to cDNA by reverse transcriptase after binding a synthetic cDNA primer. The RNA strand of the RNA: DNA hybrid is digested by RNase H, resulting in single-stranded cDNA. The cDNA can then ligate to the protopromoter to initiate another cycle of amplification. An RNA target can be amplified beginning at the reverse transcription step.

ANTIBODY-CAPTURE SOLUTION HYBRIDIZATION

Antibody-capture solution hybridization is a rapid nonradioactive method for detecting target RNA. The method uses hybridization of a DNA probe to an RNA target in solution, followed by separation of RNA : DNA hybrids from unhybridized probe onto a suitable solid

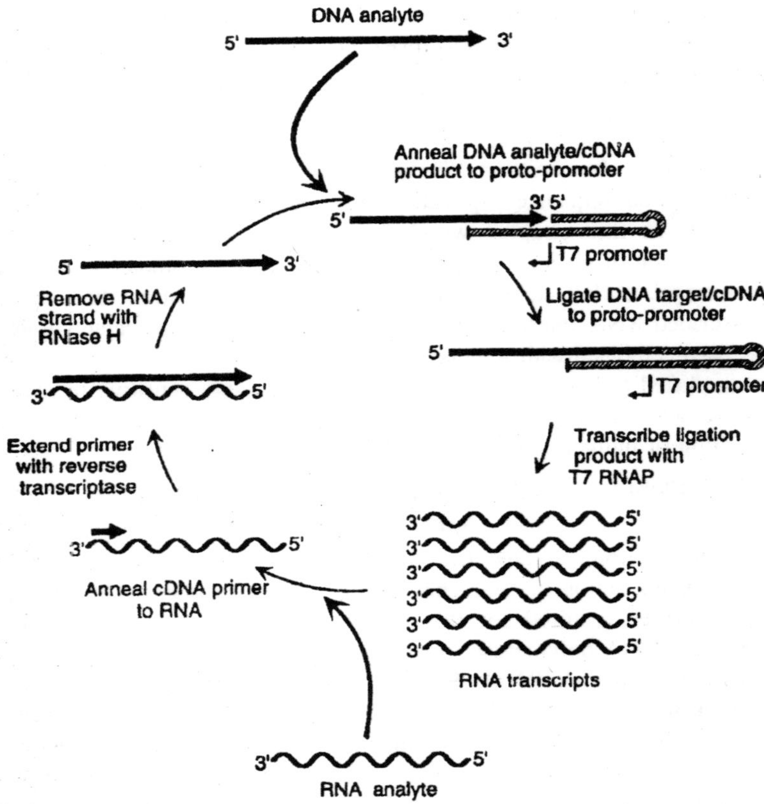

Fig. 13.1. Schematic representation of LAT amplification.

support. RNA : DNA hybrids are captured using magnetic beads carrying immobilized antibody against RNA : DNA hybrids. The enzyme-labeled probe can be detected using an appropriate colorimetric, fluorescent, or chemiluminescent substrate. The probes were designed to hybridize to the RNA product of the LAT amplification reaction because RNA is the predominant product of the amplification.

Protocols

For routine laboratory procedures, a reagent mix should be prepared that contains the required buffers, ribo- and deoxyribonucleoside triphosphate substrates, cDNA primer, and proto-promoter for amplification of specific RNA or DNA sequences. The required enzymes—T7 RNA polymerase, T4 DNA ligase, Superscript II reverse transcriptase, and *Escherichia coli* RNase H (GILBO/ BRL)—should be added as a mixture from a concentrated preparation. Enzyme and reaction mixes are stable for at least 1 month when stored at −20°C. Amplification reactions are assembled by adding the reagent mix, a ribonuclease inhibitor, and the enzyme mix to produce a master mix that is aliquoted for individual reactions. LAT amplification reactions are typically performed in 25-μl volumes.

Stock Solutions

All stock solutions must be prepared as sterile RNase-free solutions. Where appropriate, solutions may be treated with diethylpyrocarbonate (DEPC) to inactive RNases. It is sufficient to prepare solutions using virgin chemicals and plasticware with DEPC-treated sterile distilled water, followed by 0.2-μm filtration.

LAT amplification is extremely sensitive to the activity of each enzyme in the reaction. Unit values given for commercially available enzyme preparations are often inaccurate and cannot be used to determine the amounts of each enzyme to add to the mixture. Enzymes were obtained as concentrated preparations from GIBCO/BRL.

Enzymes

1. T7 RNA polymerase, 12,000 U/μl
2. Superscript II reverse transcriptase, 1050 U/μl
3. T4 DNA ligase, 16.2 U/μl
4. *E.coli* RNase H, 1 U/μl

Reagents

1. 2 M potassium glutamate
2. 2 *M* Tris-HCl, pH 8.3
3. 1 *M* $MgCl_2$
4. 40% (w/v) polyethylene glycol (PEG) 8000
5. 10% Triton X-100
6. 1 *M* dithiothreitol (DTT)
7. Nucleotide mix: 3.125 mM each dNTP, 12.5 mM each NTP; prepared from 100 mM stock solutions using DEPC-treated sterile water
8. 5× Reaction buffer: 0.25 *M* Tris-HCl, pH 8.3, containing 0.875 *M* potassium glutamate, 30 mM $MgCl_2$, 0.05% Triton ×-100, and 50 mM DTT
9. Enzyme storage buffer: 20 mM Tris-HCl, pH 7.7, 100 mM potassium glutamate, 0.1 mM EDTA, 1 mM DTT, 0.01% Triton ×-100, 50% glycerol

Reagent mix (sufficient for 100 reactions)

1. 5× reaction buffer, 500 μl
2. Nucleotide mix, 200 μl
3. Proto-promoter oligo, 400 pmol
4. cDNA primer, 1250 pmol
5. 40% PEG-8000, 375 μl
6. DEPC-treated water, 1550 μl, final volume

Enzyme mix (final volume, 400.0 μl; sufficient for 100 reactions)

1. T7 RNA polymerase (12,000 U/μl), 3.3 μl (400 U/25-μl reaction)
2. Superscript II RT (1050 U/μl), 9.5 μl (100 U/25-μl reaction)
3. RNase H (1 U/μl), 5.0 μl (0.05 U/25-μl reaction)
4. T4 DNA ligase (16.2 U/μl), 6.2 μl (1.0 U/25-μl reaction)
5. Enzyme storage buffer, 376.0 μl

   Note: Use 4 μL enzyme mix per 25-μl reaction.

Oligonucleotides

Oligonulceotides were synthesized by phosphoramidite chemistry using an Applied Biosystems Model 380A synthesizer. Primers anu oligonucleotides were purified by anion exchange HPLC or denaturing acrylamide gel electrophoresis.

Preparation of proto-promoter

The proto-promoter is a unique T7 promoter-containing oligonucleotide that has two distinct regions. The 5' region is composed of a double-stranded T7 RNA polymerase promoter sequence bridged by a hairpin loop. The 3' region is single-stranded and is complementary to the 3' end of the cDNA synthesized during the amplification process. The 3' residue of the protopromoter is cordycepin (3'-deoxyadenosine 5'-monophosphate) and therefore cannot function as a primer.

We have synthesized proto-promoter using a single 82-bp oligonucleotide or using a ligation method to assemble the proto-promoter from two smaller oligonucleotides. Although both methods have produced functional protopromoter, our best LAT amplification results were obtained using the ligation protocol. The ligation method offers an added benefit because a single generic T7 promoter oligonucleotide can be used to construct a variety of sequence- specific proto-promoters, thus saving time and reducing the cost of oligonucleotide synthesis.

Oligonucleotides 1123 and 742 were synthesized from cordycepin CPG. The 5' termini of oligonucleotides used for proto-promoter construction were chemically phosphorylated using 5' Phosphate-ON

CE-Phosphoramidite or treated with T4 polynucleotide kinase (GIBCO/BRL) in an excess of ATP). For ligation we used 6 nmol each of oligo 742 and 1124. Ligated product was electrophoresed on an 8 *M* urea, 8% polyacrylamide gel. Proto-promoter product was excised from the gel and eluted with 0.5 *M* NH_4OAc, 0.1% SDS, 10 mM Tris-HC1, pH 8.0, 1 mM EDTA. Approximately 80-90% of the two oligos ligated. Gel-purified proto-promoter was desalted by Sephadex G-50 (Pharmacia) chromatography, eluted with DEPC-treated sterile water, and then dried. Proto-promoter was resuspended in 200 μl DEPC-treated sterile water and quantified by A_{260} determination.

The single-step synthesis proto-promoter, oligo 1123, used cordycepin CPG and was chemically phosphorylated with 5' Phosphate-ON CE-Phosphoramidite. Material from a crude synthesis was gel purified as described.

Proto-promoter ligation protocol

Materials

1. Sterile distilled water
2. 5× Ligase/buffer (GIBCO/BRL): 0.25 *M* Tris-HCl pH 7.6, 50 mM $MgCl_2$, 5 mM ATP, 5 mM DTT, 25% (w/v) PEG-8000
3. Generic T7 promoter oligo (742, 208 *μM)*
4. Sequence-specific "top-strand" oligo (732, 83.5 *μM)*
5. T4 polynucleotide kinase
6. T4 DNA ligase
7. 0.5 *M* EDTA, pH 8.0
8. Formamide

Procedure

1. To two separate 1.5-ml microfuge tubes, add:

Component	*A*	*B*
Sterile distilled water	43.7 μl	—
5X ligase buffer	20.0 μl	20.0 μl
6 nmol oligonucleotide 742	28.8 μl	—
6 nmol oligonucleotide 732	—	72.5 μl
T4 polynucleotide kinase (10 U/μl)	2.5 μl	2.5 μl
Final volume	95.0 μl	95.0 μl

2. Incubate both reactions 5 min at 37°C.
3. Incubate both reactions 5 min at 94°C to heat inactivate the kinase.
4. Transfer the contents of both reactions to a single 1.5-ml tube.
5. Add 10 μl T4 DNA ligase (1 U/μl).

6. Incubate 1 hr at 37°C.
7. Add 10 μl 0.5 *M* EDTA to stop the reaction.
8. Lyophilize in a Speed-Vac.
9. Resuspend concentrated ligation product with 50 μl deionized formamide.
10. Electrophorese the ligation products on a denaturing 8% polyacrylamide gel according to standard protocols..
11. Visualize proto-promoter product by UV shadowing.
12. Excise proto-promoter band from the gel. Crush the gel matrix and elute with 2-3 volumes 0.5 *M* NH_4OAc, 0.1% SDS, 10 mM Tris-HC1, pH 8.0, 1 mM EDTA at 37°C for 16 hr.
13. Desalt by gel-exclusion chromatography (PD-10 column; Pharmacia), eluting with DEPC-treated sterile water.
14. Lyophilize as in Step 8.
15. Resuspend proto-promoter in 200 μl DEPC-treated sterile water and quantify by A_{260} determination.

LAT Amplification

1. Prepare a master mix by combining 15.5 μl reagent mix, 0.5 μl placental ribonuclease inhibitor, and 4 μl enzyme mix for each reaction. Always make sufficient mix for one additional reaction, for example, if 20 reactions are to be performed, prepared a master mix sufficient for 21 reactions. To a sterile RNase-free 1.5-ml microfuge tube on ice, add in order:

325.5 μl	reagent mix
10.5 μl	RNasin (10 U/μl)
84.0 μl	enzyme mix

for a final volume of 420.0 μl.

Note: The requirement for placental ribonuclease inhibitor is dependent on the quality of the analyte sample. In "clean" model systems, RNase inhibitor is not needed; however, LAT amplification of material isolated from clinical specimens requires RNasin. Exercise caution when using ribonuclease inhibitors. Add RNasin to the reaction mix only after addition of reducing agent (DTT). RNasin requires sulfhydryl reagents (1 mM DTT) for activity and is irreversibly denatured in the absence of these agents. Some preparations of ribonuclease inhibitor include inactivated RNases that copurify with the inhibitor. Treatments that denature the inhibitor (repeated freeze-thawing, heat, absence of DTT, etc.)

can release these RNases, which can degrade RNA in subsequent procedures.

2. Mix the solution thoroughly by gently vortexing; then place on ice.
3. Transfer 20-μl aliquots of the master mix into sterile 0.5-ml microfuge tubes on ice.
4. Add 5 μl test sample. (Typical levels of analyte assayed contain 100-10,000 copies. The final reaction composition is 50 mM Tris-HCl, pH 8.3, 175 mM potassium glutamate, 6% PEG 8000, 6 mM $MgCl_2$, 10 mM DTT, 0.01% Triton ×-100, 0.25 mM each dNTP, 1 mM each NTP, 0.5 μM cDNA primer, 0.15-0.2 μM T7 promoter oligonucleotide, 16 U/μl T7 RNA polymerase, 4 U/μl Superscript II reverse transcriptase. 0.04 U/μl T4 DNA ligase, 0.002 U/μl *E. coli* RNase H, 0.2 U/μl RNase inhibitor, 100-10,000 copies target sequence.)
5. Mix and incubate at 37°C for 3 hr.
6. Add 25 μl 50 mM EDTA to terminate the reaction.

Detection of Amplification Products

Amplification products can be detected by a variety of nucleic acid hybridization techniques. We have used a modification of the dot-bolt filter hybridization method of Thomas (1983) using a ^{32}P-labeled oligonucleotide probe that is specific for either the RNA or the cDNA synthesized during the reaction. Additionally, we have developed a chemiluminescent immunocapture assay for the direct detection of RNA amplification products, employing a monoclonal antibody specific for RNA : DNA heteroduplex molecules and an alkaline phosphatase-labeled oligonucleotide probe.

Comparison of relative signal intensities between reaction products and dilutions of the original analyte (hybridization standards) provides a measure of the level of amplification attained from the LAT process.

Membrane hybridizations

^{32}P-labeling of oligonucleotide probe

1. To a 0.5-ml microfuge tube on ice, add:
   - 2.0 μl 5× Forward buffer [300 mM Tris-HCl, pH 7.8, 75 mM 2-mercaptoethanol (BME), 50 mM $MgCl_2$, 1.65 *μM* ATP]
   - 2.0 μl oligonucleotide probe (5 pmol/μl)
   - 5.0 μl [γ-^{32}P]ATP (10 mCi/ml; 3000 Ci/mmol)
   - 0.5 μl T4 polynucleotide kinase (10 units/μl)

   for a final volume of 10.0 ktl.

2. Incubate at 37°C for 10 min.
3. Terminate reactions by adding EDTA to 20 mM final concentration.
4. Remove unincorporated ATP from radiolabeled probe using any appropriate methodology (gel exclusion chromatography, ethanol precipitation, ion-exchange chromatography, etc.).

Glyoxal dot-blot procedure

- Materials
  1. Glyoxal solution: 1 *M* glyoxal, 50% dimethylsulfoxide (DMSO), 50 mM $NaPO_4$, pH 7.0
     - 1 volume deionized glyoxal (40%)
     - 2 volumes DMSO
     - 1 volume 200 mM sodium phosphate, pH 6.8-7.0
  2. Hybridization solution: 1 *M* NaCl, 30% formamide, 5% SDS, 50 mM $NaPo_4$, pH 7.4, 1 mM EDTA, 0.1% gelatin, 50 μg/ml yeast tRNA
- Protocol
  1. Combine 5μl LAT amplification reaction with 25 μl glyoxal solution.
  2. Incubate at 50°C for 1 hr, then place on ice.

     Note: In addition to LAT reaction products, dilutions of the RNA or cDNA analyte can be treated with glyoxal and used as hybridization standards.
  3. Add 150 μl 20x SSC (3 *M* NaCl, 0.3 *M* sodium citrate, pH 7.0) to each glyoxal reaction. Mix and apply the entire volume to a suitable hybridization membrane with a microfiltration manifold according to standard protocols. We recommend using Biodyne B nylon membrane (GIBCO/BRL).
  4. Bake the membrane at 80°C for a minimum of 30 min to remove glyoxal adducts from the target and fix the target to the membrane. Because of the small size (~100 bases) of the LAT amplification products, membranes should also be fixed by UV treatment. Place the membrane on a 302-nm UV transilluminator (target side down) and expose for 2 min.
  5. Place the filter in a heat-sealable plastic bag and prehybridize for 30 min. Use approximately 1 ml hybridization solution per 20 cm^2 membrane.
  6. Replace the prehybridization solution with an equivalent volume of fresh hybridization solution containing 1×10^7 dpm/ml ^{32}P-

labeled oligonucleotide probe (specific activity = 1 × 10^8 dpm/μg). Reseal the bag and hybridize overnight (16 hr).

7. Wash the membrane 5 times for 5 min each with 0.1× SSC, 0.1% SDS at 42°C.
8. Detect hybridized probe by phosphoimaging or autoradiography according to standard protocols.

For detection of HPV16 sequences, prehybridization/hybridization was performed at 42°C when using the RNA probe oligo (5'-GTATATTGTGGAGACCCTGGAACTATAGG-3') and at 37°C when using the cDNA probe oligo (5'-TCCTATAGTTCCAGGGTCTCCACAATATAC-3'). Hybridization and wash conditions should be optimized for each individual probe.

Immunocapture assay

Preparation of anti-RNA : DNA-coated paramagnetic beads

Mouse monoclonal IgG specific for RNA : DNA heteroduplex (anti-RNA : DNA) were prepared essentially as described by Boguslawski *et al.* (1986). A high affinity IgG2a isolate was used for covalent coupling to magnetic polystyrene beads, essentially as recommended by the manufacturer. Uncoated M-450 magnetic beads require activation with *p*-toluenesulfonyl chloride before use. Optionally, tosyl-activated magnetic beads may be obtained from Dynal. Activated beads are used for direct covalent coupling of the primary monoclonal antibody.

- Materials

1. Tosyl-activated magnetic beads (Dynabeads)
2. Anti-RNA : DNA, purified mouse monoclonal antibody (mg/ml in 50% glycerol and water)
3. Borate buffer: 0.2 *M* $Na_2B_4O_7 \cdot 10\ H_2O$ (Borax), pH 9.2
4. 1.0 *M* ethanolamine-HCl, pH 9.5
5. Tween20®
6. 0.1 *M* phosphate buffered saline (PBS), pH 7.2
7. Tris-buffered saline (TBS): 0.05 *M* Tris-HC1, pH 7.5, 0.1 *M* NaC1, 0.1% BSA, 0.01% thimerosal
8. TBS with Tween 20®: 0.05 *M* Tris-HCl, pH 7.5, 0.1 *M* NaCl, 0.1% BSA, 0.01% thimerosal, 0.1% Tween20®

- Protocol

1. Rinse tosyl-activated M-450 beads once with sterile distilled water and resuspend at a final concentration of 30 mg/ml.

2. Dilute anti-RNA : DNA in 0.2 *M* borate solution, pH 9.5, to a final concentration of 150 μg/ml.
3. Add equal volumes of activated beads and antibody solution to a 15-ml screw-cap polypropylene tube. Cap and seal the tube with parafilm.
4. Couple antibody to beads for 24 hr at room temperature with continuous mixing (end-over-end rotation).
5. Collect the beads with a magnetic concentrator (Dynal MPC or equivalent) and discard the supernatant.
6. Wash antibody-coupled beads with 5 ml 0.1 *M* PBS for 10 min using end-over-end rotation. Collect the beads with magnetic concentration and discard the supernatant.
7. Add 5 ml 1 M ethanolamine-HCl, pH 9.5, 0.1% Tween20. Incubate for 2 hr at room temperature with continuous mixing to block free tosyl groups. *Note:* Add Tween20 to the ethanolamine-HCl solution immediately before use.
8. Wash with 5 ml TBS with Tween20 for 12 hr at room temperature (end-over-end rotation).
9. Wash with 5 ml TBS for 2 hr at room temperature.
10. Collect the antibody-coated beads with a magnetic concentrator. Discard the supernatant and resuspend in TBS at a final concentration of 30 mg/ml. Store at 4°C.

    Note: Antibody-coated beads are stable for 6 mo at 4°C. Before each use, wash the beads with TBS, 0.1% Tween20 for 5 min at room temperature. Resuspend in TBS with Tween20 for a final concentration of 30 mg/ml.

Preparation of alkaline phosphatase-coupled oligonucleotide probe

A 5' Aminoalkyl-oligonucleotide specific for RNA products from LAT amplification of HPV 16 (5'-NH_2-GTATATTGTGGAGACCC-TGGAACTATAGG-3') was synthesized by standard protocols using Aminolink-2 phosphoramidite. Direct enzyme-labeled probe was prepared using the ACES Labeling System (GIBCO/BRL) according to the manufacturer's recommendations. Briefly, 20 nmol 5'-aminoalkyl oligonucleotide was activated by thiolation and purified by gel filtration. Maleimidoderivatized alkaline phosphatase was reacted with an excess of thiolated oligonucleotide to form a covalent adduct. Alkaline phosphatase-labeled oligonucleotide was purified from unconjugated oligonucleotide by gel filtration. Enzyme-linked probe is stable for up to 6 mo when stored at 4°C in 50 *mM* Tris-HCl, pH 7.5, 0.1% BSA,

0.2% sodium azide. For prolonged storage, probe should be aliquoted and stored at −20°C. Repeated freeze-thawing should be avoided since it may impair enzyme activity.

Solution hybridization and immunocapture

- Materials
  1. 45% polyacrylic acid (sodium salt)
  2. RNase A (Sigma)
  3. 2× Hybridization solution: 6× SSC, 0.5% Tween20®, 4% sodium poly acrylate, 200 μg/ml sheared denatured herring sperm DNA
  3. TNT buffer: 0.1 *M* Tris-HCl, pH 7.5, 0.6 *M* NaCl, 0.25% Tween20
  4. TE buffer: 10 mM Tris-HCl, pH 7.5, 1 mM EDTA
  5. Enzyme-linked oligonucleotide probe: ~1.5 μM Tris-HCl, pH 7.5, 0.1% BSA, 0.2% sodium azide
  6. Lumi-phos 530 reagent (GIBCO/BRL)
  7. Control RNA analyte (prepared by *in vitro* transcription of cloned target sequence)
- Protocol
  1. Dilute alkaline phosphatase-lnked oligonucleotide probe in TE buffer to a final concentration of 25nM.
  2. Prepare a master mix hybridization cocktail by combining 25 μl 2× hybridization solution and 20 μl 25 nM oligonucleotide probe/TE buffer solution for each hybridization. Prepare sufficient master mix to accommodate each LAT reaction as well as dilutions of an RNA control analyte (used to generate a standard curve).
  3. Dispense 45-μl aliquots into 1.5-ml microfuge tubes.
  4. Add 5 μl RNA sample (LAT reaction aliquot or control RNA) to each hybridization.
  5. Incubate 30 min at 42°C.
  6. Add 100 μl TNT containing 20 μg/ml RNase A and 5 μl antibody-coated beads to each hybridization reaction.
  7. Mix; then incubate 30 min at room temperature.
  8. Wash beads 6 times with 0.5 ml TNT. (a) Place tube in magnetic rack. (b) Allow beads to migrate to tube walls. (c) With tube in magnetic rack, remove supernatant by aspiration or decanting.

Characteristics of the LAT Amplification System

The LAT amplification is an isothermal system designed for exponential amplification of nucleic acids. The system is capable of amplifying RNA or DNA. LAT amplification produces both RNA and single-stranded DNA, but RNA products are produced in larger quantities than DNA products. An overall amplification of 10^7- to 10^8-fold is usually achieved within 3 hr.

A model system based on the sequences of HPV16 has been used to characterize LAT amplification and its products. The model system consisted of a 660-base *in vitro* transcript produced from a cloned fragment of the HPV16 genome. The system was designed so the 5' end region of this RNA template would be amplified. A series of different cDNA primers was used to generate cDNAs with different lengths that were subsequently amplified. The 3' ends of these cDNAs were complementary to the protopromoter, which allowed amplification of all cDNAs using the same reagents.

A typical amplification reaction results in production of both RNA and DNA products. Quantitative hybridization of probes has shown that 5- to 10-fold more RNA is produced than DNA. The full-length transcript is the major RNA product, RNAse H degradation products make up the smaller species. The DNA products are of two species: (1) the full-length cDNA corresponding to the reverse transcript of the RNA produced in the reaction and (2) a slightly larger species produced as a result of ligation between cDNA and the proto-promoter.

We studied the kinetics of LAT amplification using a 119-base region of the *in vitro* RNA transcript just described. Amplification was shown to be biphasic. An exponential amplification occurred within 2-3 hr that produced an accumulation of approximately 10^{11} molecules of RNA. The subsequent decline in reaction rate is probably due to exhaustion of the reaction components. In other experiments, we have shown that the kinetics of amplification vary with the length of the product to be amplified; shorter fragments are amplified more rapidly.

Amplification of the DNA templates by LAT requires the presence of an available 3' end. For RNA targets, this is accomplished by synthesis of a cDNA exploiting a discrete 5' mRNA terminus. For DNA templates, this 3' end can be obtained by digesting the DNA template with an appropriate restriction enzyme. This hypothesis was tested using a clone containing the genome of HPV16. The plasmid was digested with *Alμl*, which generated DNA restriction fragments with a 3' end region complementary to the proto-promoter used in LAT amplification. This allowed amplification of the DNA templates using the same reagents used for amplification of RNA. Amplification of DNA templates was demonstrated to be equally efficient to that of RNA templates. We also demonstrated that DNA amplification was

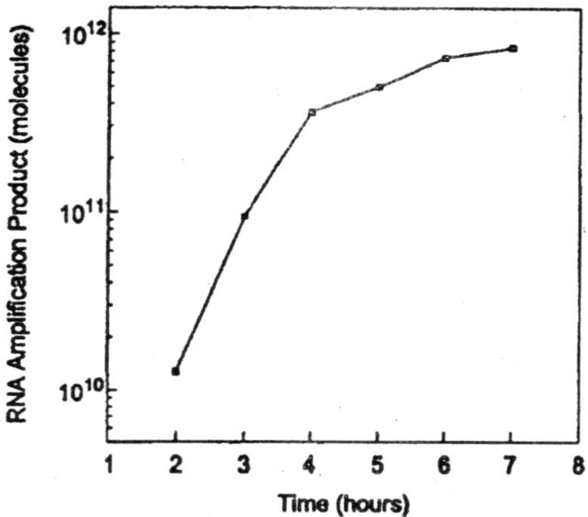

Fig. 13.2. Time course of LAT amplification.

not only dependent on the generation of the 3' end, but also on denaturation of DNA prior to starting the LAT reaction; double-stranded DNA templates fail to amplify. This feature provides a very selective procedure for specific amplification of RNA, even in the presence of DNA. Amplification of RNA may be important in some diagnostic procedures where expression of certain genes is more important than their mere presence. This may prove to be the case with HPV16, since the expression of transcripts derived from the E6 region has been implicated in tumorigenicity.

Chemiluminescence Detection of Amplified Products

A nonradioactive chemiluminescence assay was developed for rapid detection of the amplified RNA from LAT. The assay used a synthetic oligonucleotide probe labeled with alkaline phosphatase. This probe is complementary to the RNA product of amplification and is used to detect and quantify the levels of RNA generated. The products of amplification reactions were mixed and hybridized in solution to the oligonucleotide probe. The RNA : DNA hybrids formed as a result of hybridization were then captured using magnetic beads. The activity of the alkaline phosphatase associated with the hybrids was detected using the chemilumine scent substrate Lumiphos 530 [4-methoxy4-(3-phosphate-phenol)spiro(1,2-dioxetane-3,2'-adamantane]. Use of a known amount of RNA target and the subsequent generation of a standard curve allowed the quantification of the RNA products generated by amplification. A minimum of $1 \times : 10^8$ copies of target molecules can be detected using this hybridization procedure and a single

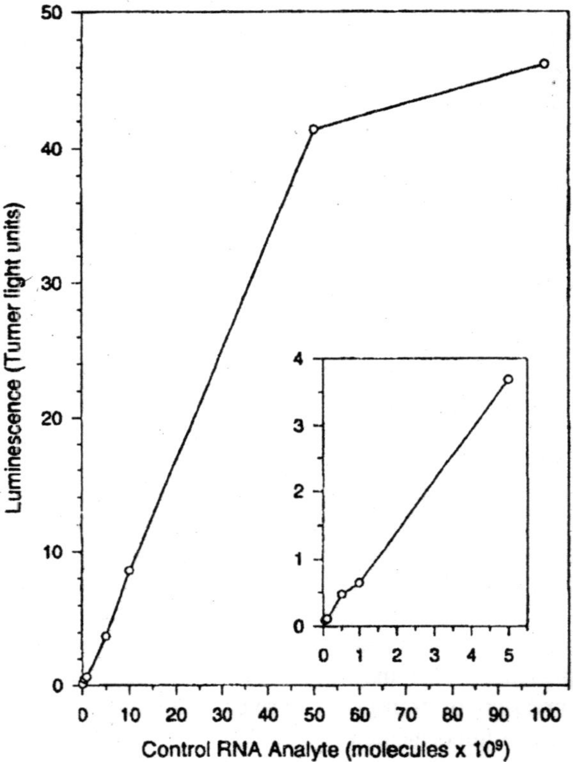

Fig. 13.3. Chemiluminescent detection of HPV16 RNA.

oligonucleotide probe. The assay is linear up to 5×10^{10} molecules. These limits provide a useful range for the quantification of LAT amplification products. LAT amplification was capable of amplifying as few as 10 copies of RNA target. The extend of amplification was relatively linear and the amount of amplified product correlated with the amount of input target.

In other experiments, we compared the accuracy of RNA quantification with dot-blot hybridization using radioactivity labeled oligonucleotide probes. These studies have shown equal sensitivity for both systems, as well as excellent agreement between results.

Amplification and Detection of HPV in Clinical Specimens

LAT amplification was applied to detection of HPV 16 RNA in clinical specimens. The RNA molecule targeted for these assays was the major E6/E7 spliced transcript, E6* I, which is believed to be the primary molecular mechanism for expression of the oncogenic protein E7 in HPV16. LAT amplification and RT-PCR were performed using total RNA isolated from 10 human cervical carcinoma biopsy samples; SiHa (HTB 35) cells, an HPV16-transformed cell line known to contain E6* mRNA, and C-33A (HTB 31) cells, a cell line derived from cervical cancer biopsies and devoid of HPV, as a negative control. The 5' terminus of the E6/E7 mRNAs initiated at the P97 promoter was targeted for amplification by both methods using oligonucleotide constructs that targeted identical sequence regions. An oligonucleotide probe designed to span the E6* I splice donor/acceptor sites was used for specific detection of the E6* I amplification product.

First strand synthesis was primed using primer B (TGCTTTTC-TTCAGGACACAG), designed to be complementary to a region immediately 3' to the HPV16 E6/E7 major splice acceptor site (nucleotides 431-450). Approximately one-fourth of each cDNA synthesis reaction was used for PCR amplification (100-μl volumes) in conjunction with primer A (GAGAACTGCAATGTTTCAGG), which was specific for the 5' terminus of E6/E7 transcripts (homologous to the region from nucleotide 94 to 113 of the HPV 16 genome). PCR amplification was carried through 25 temperature cycles according to standard protocols. The 176-bp E6* I amplification product was detected by Southern blot using a ^{32}P-labeled oligonucleotide probe (TGACAGTT-AATACACCTCACG) that spanned the splice donor and acceptor sites of the E6* I transcript.

LAT amplification reactions were performed with 5 μl total RNA sample in a final volume of 25 μl, as described in this chapter, using primer B as the cDNA primer and a proto-promoter containing a 3' single-stranded overhang homologous to the 5' terminus of HPV16 E6/E7 transcripts (nucleotides 94 to 113). Reactions included 20 U RNasin to suppress endogenous RNase activities that may have persisted in the RNA isolation procedure. Reactions were incubated for 4 hr at 37°C, and E6* I amplification product was detected by glyoxal dot blot using conditions identical to those for PCR product Southern blot. Although relative signal intensities between LAT and PCR amplification product for a given sample sometimes differ, an excellent correlation is observed between the two methods for detection of E6* I amplification products.

14

VIRAL DIVERSITY

PARVOVIRUSES

Parvoviruses are amongst the smallest known viruses, with virions in the range 18–26 nm in diameter. They derive their name from the Latin *parvus* (= small). The family *Parvoviridae* has been divided into two subfamilies: the *Parvovirinae* (vertebrate viruses) and the *Densovirinae* (invertebrate viruses). The subfamily *Parvovirinae* includes the genus *Dependovirus*, the members of which are defective, normally replicating only when the cell is co-infected with a helper virus. Other parvoviruses that do not require helper viruses are known as autonomous parvoviruses.

Examples of Parvoviruses

Dependoviruses

The first dependovirus to be discovered was observed in the electron microscope as a contaminant of an adenovirus preparation. The contaminant turned out to be a satellite virus: a defective virus dependent on the help of the adenovirus for replication. The satellite virus was therefore called an adeno-associated virus. Other dependoviruses (distinct serotypes) have since been found in adenovirus preparations and in infected humans and other species. Results of surveys using serological methods and PCR detection of virus DNA indicate that dependovirus infections are widespread.

Not all dependoviruses have an absolute requirement for the help of an adenovirus. Other DNA viruses (e.g. herpesviruses) may sometimes act as helpers, and some dependoviruses may replicate in the absence of a helper virus under certain circumstances.

Dependoviruses are valuable gene vectors. They are used to introduce genes into cell cultures for mass-production of the proteins encoded by those genes, and they are being investigated as possible vectors to introduce genes into the cells of patients for the treatment of various genetic diseases and cancers. One of the advantages of dependoviruses for such applications is the fact that they are not known to cause any disease, in contrast to other viruses under investigation, such as retroviruses.

Autonomous parvoviruses

A parvovirus that does not require a helper virus was discovered in serum from a healthy blood donor. The virus, named after a batch of blood labelled B19, infects red blood cell precursors. Many infections with B19 are without signs or symptoms, but some result in disease, such as fifth disease (erythema infectiosum), in which affected children develop a 'slapped-cheek' appearance.

Other diseases caused by B 19 virus include

1. acute arthritis
2. aplastic anaemia in persons with chronic haemolytic anaemia
3. hydrops foetalis (infection may be transmitted from a pregnant woman to the foetus and may kill the foetus).

In 2005 a new human parvovirus was discovered using a technique for molecular screening of nasopharyngeal aspirates from children with lower respiratory tract disease. The virus is related to known parvoviruses in the genus *Bocavirus*.

Viruses in the subfamily *Densovirinae* cause the formation of dense inclusions in the nucleus of the infected cell. Some of these viruses are pathogens of the silkworm (*Bombyx mori*), and can cause economic damage to the silk industry.

Parvovirus Virion

Parvoviruses are small viruses of simple structure with the ssDNA genome enclosed within a capsid that has icosahedral symmetry.

Capsid

The parvovirus capsid has icosahedral symmetry and is built from 60 protein molecules. One protein species forms the majority of the capsid structure and there are small amounts of between one and three other protein species, depending on the virus. The proteins are numbered in order of size, with VP1 the largest; the smaller proteins are shorter versions of VP1. Each protein species contains an eight-stranded β barrel structure that is common to many viral capsid proteins, including those of the picornaviruses. The virion is roughly spherical, with surface protrusions and canyons. At each of the vertices of the icosahedron there is a protrusion with a pore at the centre.

Genome

Parvoviruses have genomes composed of linear ssDNA in the size range 4–6 kb. At each end of a DNA molecule there are a number of short complementary sequences that can base pair to form a secondary structure. Some parvovirus genomes have sequences at their ends known as inverted terminal repeats (ITRs), where the sequence at one end is complementary to, and in the opposite orientation to, the sequence at the other end. As the sequences are complementary, the ends have identical secondary structures. Other parvoviruses have a unique sequence, and therefore a unique secondary structure, at each end of the DNA.

During replication, parvoviruses with ITRs generate and package equal numbers of (+) and (H) strands of DNA, while most viruses with unique sequences at the termini do not. The percentages of virions containing (+) DNA and (H) DNA therefore vary with different viruses. In a (–) DNA the genes for non-structural proteins are towards the 3' end and the structural protein genes are towards the 5' end.

Parvovirus Replication

The small genome of a parvovirus can encode only a few proteins, so the virus depends on its host cell (or another virus) to provide important proteins. Some of these cell proteins (a DNA polymerase and other proteins involved in DNA replication) are available only during the S phase of the cell cycle, when DNA synthesis takes place.

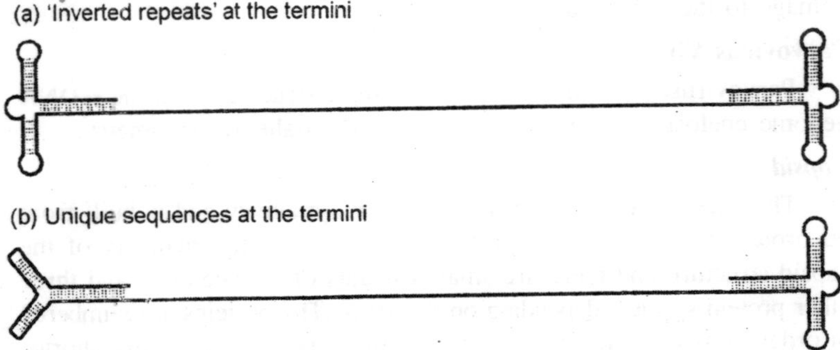

Fig. 14.1. Base pairing at the termini of parvovirus DNA.

This restricts the opportunity for parvovirus replication to the S phase. Contrast this situation with that of the large DNA viruses, such as the herpesviruses, which encode their own DNA-replicating enzymes, allowing them to replicate in any phase of the cell cycle.

Attachment and entry

A virion attaches to receptors on the surface of a potential host cell. In the case of B19 virus the host cell is a red blood cell precursor and the receptor is the blood group P antigen. The virion enters the cell by endocytosis and is released from the endosome into the cytoplasm, where it associates with microtubules and is transported to a nuclear pore. With a diameter of 18–26 nm, the parvovirus virion is small enough to pass through a nuclear pore, unlike the herpesvirus nucleocapsid, though there is evidence that the virion must undergo some structural changes before it can be transported into the nucleus. Nuclear localization signals have been found in the capsid proteins of some parvoviruses.

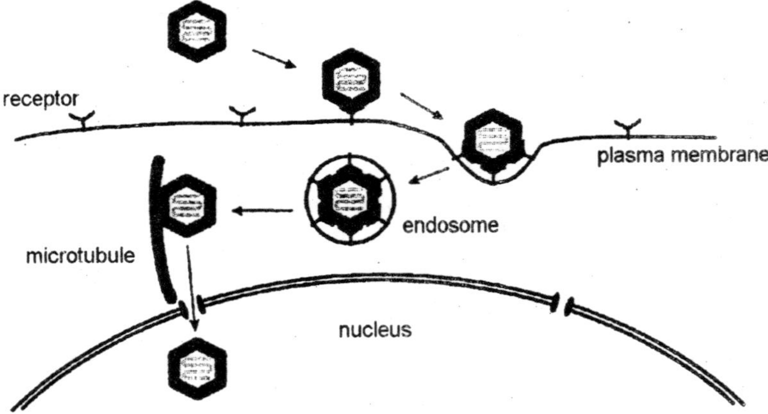

Fig. 14.2. Parvovirus attachment and entry.

Single-stranded DNA to double-stranded DNA

In the nucleus the single-stranded virus genome is converted to dsDNA by a cell DNA polymerase. The ends of the genome are double stranded as a result of base pairing, and at the 3' end the -OH group acts as a primer to which the enzyme binds.

Transcription and translation

The cell RNA polymerase II transcribes the virus genes and cell transcription factors play key roles. The primary transcript(s) undergo various splicing events to produce two size classes of mRNA. The larger mRNAs encode the non-structural proteins and the smaller mRNAs encode the structural proteins. The non-structural proteins are phosphorylated and play roles in the control of gene expression and in DNA replication.

DNA replication

After and virion assembly conversion of the ssDNA genome to dsDNA, the DNA is replicated by a mechanism called rolling-hairpin replication. This is a leading strand mechanism and sets parvoviruses apart from other DNA viruses, which replicate their genomes through leading and lagging strand synthesis.

Procapsids are constructed from the structural proteins and each is filled by a copy of the virus genome, either a (+) DNA or a (–) DNA as appropriate. One of the non-structural proteins functions as a helicase to unwind the dsDNA so that a single strand can enter the procapsid.

Dependovirus replication

When a cell is co-infected with a dependovirus and an appropriate helper virus there is a productive infection with both viruses.

In nature it is perhaps more common for a dependovirus to infect a cell in the absence of a helper virus, in which case the virus genome, after it has been converted to dsDNA, may be integrated into a cell chromosome. Integration occurs as a result of recombination between the cell and viral DNAs and results in a latent infection. In human cells the virus DNA is integrated at a specific site in chromosome 19. Latent dependovirus infections have been found in a number of cell lines of human and monkey origin. If a cell with a latent dependoviral genome becomes infected with an appropriate helper virus then a productive infection with both viruses can ensue.

Other ssDNA viruses

These viruses, and indeed the majority of the known ssDNA viruses, have circular genomes. The only viruses known with ssDNA linear genomes are the parvoviruses.

Table 14.1. Examples of ssDNA viruses

	Family	*Example*
Animal viruses	*Circoviridae*	Porcine circovirus
Plant viruses	*Geminiviridae*	Maize streak virus
Bacterial viruses	*Microviridae*	Phage φX174

Reoviruses

Icosahedral viruses with dsRNA genomes isolated from the respiratory tracts and enteric tracts of humans and animals, and with which no disease could be associated (orphan), became known as *reoviruses*. A large number of similar viruses have been found in mammals, birds, fish, invertebrates including insects, plants and fungi. Many of these viruses are causative agents of disease, but the original name has been preserved in the family name *Reoviridae*, and has been incorporated into the names of several of the genera within the family.

Table 14.2 Some of the genera in the family *Reoviridae*

Genus	*Hosts*	*Number of dsRNA segments*
Cypovirus	Insects	10
Orbivirus	Mammals, birds, invertebrates	10
Orthoreovirus	Mammals, birds	10
Phytoreovirus	Plants, insects	12
Rotavirus	Mammals, birds	11

The original reoviruses are incorporated into the genus *Orthoreovirus*. The avian viruses are important disease agents, but most orthoreovirus infections of mammals are asymptomatic. The majority of humans become infected with orthoreoviruses early in life and have specific serum antibodies by early adulthood.

It is interesting to note that most of the plant-infecting reoviruses are transmitted between plants by insect vectors. The viruses replicate in both the plant and the insect, generally causing disease in the plant, but little or no harm to the infected insect.

The main focus of this section is on the rotaviruses, which have been the subjects of intensive study because they are amongst the most important agents of gastroenteritis in humans and animals.

Rotavirus Virion

Rotaviruses were first described in 1963, when they were observed during electron microscopy of faecal samples from monkeys and mice. Spherical virions about 75 nm in diameter, with structures resembling the spokes of a wheel, were described, so the viruses were named after the Latin word *rota* (= wheel). Similar viruses were observed 10 years later during electron microscopy of faecal samples from children with diarrhoea.

The virion, which has icosahedral symmetry, is also known as a triple-layered particle as the capsid has three layers, each constructed from a distinct virus protein (VP). The inner and middle layers, constructed from VP2 and VP6 respectively, are perforated by channels. The middle layer contains the 'spokes' of the 'wheel' and is the major component of the virion. The outer layer is constructed from VP7, which is glycosylated. It is unusual to find a glycoprotein in a naked virion, but VP7 is associated with a membrane within the cell before it is incorporated into the virion. Three other protein species are found in the virion: VP1 and VP3 in the core, and VP4, which forms 60 spikes at the surface.

The proteins are numbered in order of their sizes. The three largest proteins are found towards the centre of the virion; within the inner capsid layer (VP2) associated with the genome are 12 copies of VP1 and VP3, which are enzymes. VP1 is the RNA-dependent RNA

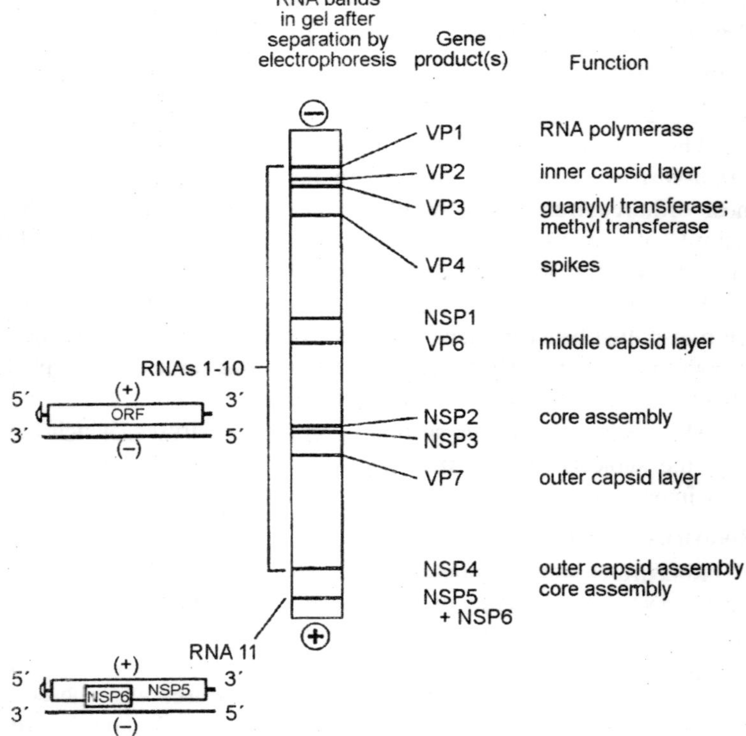

Fig. 14.3. The rotavirus genome and gene products.

polymerase while VP3 has guanylyl transferase and methyl transferase activities. There is one copy each of VP1 and VP3 attached to the inner capsid layer at each of the 12 vertices of the icosahedron.

The 11 dsRNA segments that make up the rotavirus genome can be separated according to size by electrophoresis in a sodium dodecyl sulphate-polyacrylamide gel. Each RNA segment encodes one protein, with the exception of one segment, which encodes two proteins. Hence 12 proteins are encoded: six structural proteins (VP) and six non-structural proteins (NSP). In the diagrams the RNA strands are colour-coded to distinguish the plus strand (the coding strand) from the minus strand.

Rotavirus replication

Rotaviruses infect cells called *enterocytes* at the ends of the villi (finger-like extensions) in the small intestine.

Attachment and entry

The mechanisms by which rotaviruses attach to and enter their host cells are complex and many details remain uncertain. Cleavage of the spike protein VP4 to its products VP8* and VP5* by proteolytic enzymes such as trypsin results in much more rapid entry into the cell. Binding of a virion to a cell is initially via sites on the spike proteins (VP4, VP8*, VP5*) and then via sites on the capsid surface (the glycoprotein VP7). It is likely that these proteins interact with a number of cell surface proteins; there is evidence that VP5* and VP7 bind to integrins. There are two possible ways in which a virion can enter the cell: direct penetration of the virion across the plasma membrane and endocytosis. It is thought that direct penetration is mediated by a hydrophobic region of VP5*. This region is hidden in

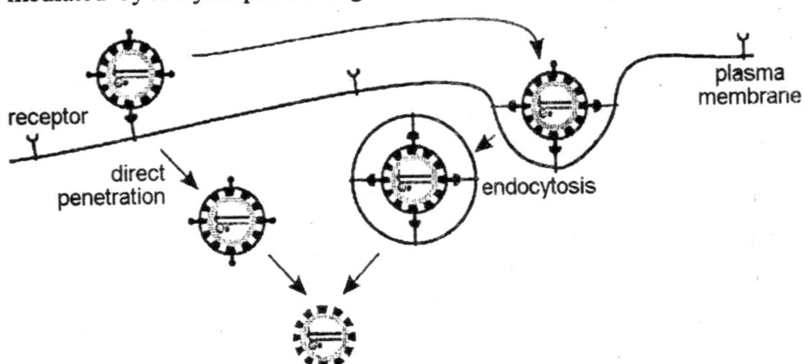

Fig. 14.4. Modes of rotavirus entry into the host cell.

the uncleaved VP4, so virions with spike proteins that have not been cleaved are unable to enter by this mechanism.

Early events

The outer layer of the virion is removed, leaving a double-layered particle in which transcription is activated. It is likely that each of the 11 genome segments is associated with a molecule of VP1, which synthesizes a new copy of the (+) RNA, and a molecule of VP3, which caps the 5' end of the new RNA. The nucleotides for RNA synthesis enter the particles through the channels in the protein layers and the transcripts are extruded from the particles through the same channels. The transcripts are not polyadenylated.

Some of the virus proteins undergo co- and post-translational modifications: VP2 and VP3 are myristylated while NSP5 is phosphorylated and O-glycosylated. The virus proteins accumulate within the cytoplasm in discrete regions known as viroplasms. There is evidence that NSP2 and NSP5 play roles in the formation of viroplasms. Transient expression of the genes for these proteins resulted in the formation of viroplasm-like structures, while reduced expression of NSP5 by RNA interference resulted in fewer and smaller viroplasms. In the viroplasms virion cores are assembled from VP1, VP2 and VP3.

Newly synthesized (+) RNAs enter the cores, and a rigorous selection procedure ensures that each core receives one each of the 11 RNA species, i.e. a full genome complement. This procedure, which operates for all viruses with multipartite genomes, involves the recognition of a unique sequence in each genome segment.

Synthesis of (-) RNA takes place during the entry of the (+) strands into the core, VP1 again acting as the RNA polymerase. The dsRNA of the infecting virion therefore remains intact and the mode of replication is conservative, in contrast to the replication of the dsRNAs of some other viruses. VP6 is added to the core, forming the second layer of the capsid. The resulting structure is a double-layered particle similar to that derived from the infecting virion.

Late events

A further round of transcription takes place within the double-layered particles. In contrast to the early transcripts, the late transcripts are not capped. The translation machinery of the cell undergoes changes that result in it selecting uncapped transcripts in preference to capped transcripts. Hence, translation of cell proteins is shut down while translation of virus proteins continues.

The virus requires different quantities of each of its 12 proteins; for example, a large amount of the main capsid protein (VP6) is required, but a relatively small amount of VP1. Although there are equimolar amounts of each of the 11 dsRNAs, the mRNA species are not made in equimolar amounts, and there are also controls at the level of translation. Hence, there are several mechanisms used by the virus to control the quantity of each protein produced.

The final stages of virion assembly involve the addition of the outer layer of the capsid and the spikes. VP7 and NSP4 are synthesized and N-glycosylated in the rough endoplasmic reticulum, where they remain localized in the membrane. NSP4 has binding sites for both VP4 and double-layered particles. After binding these components the immature virion buds through the membrane into a vesicle within the endoplasmic reticulum.

The vesicle membrane forms a temporary 'envelope' which contains VP7. Cleavage of VP7 molecules releases them from the membrane to build up the outer layer of the virion and VP4 is added to form the virion spikes. There is some uncertainty as to whether the spikes are added within the endoplasmic reticulum or at another location. Virions are released from the cell either by lysis or by exocytosis.

Rotavirus disease

Enterocytes on the villi are destroyed as a result of rotavirus infection and this leads to reduced absorption of water, salts and sugars from the gut. There is evidence that the tight junctions between cells are damaged by the non-structural protein NSP4, allowing leakage of fluid into the gut. These effects of virus infection, together with the secretion of water and solutes by secretory cells, result in diarrhoea and this can lead to dehydration. Treatment is a relatively simple matter of rehydrating the patient with a solution of salts and sugar, but the means for this are not available in some parts of the world. It estimated that each year there are around half a million deaths of infants and young children due to rotavirus infections.

Although most rotavirus infections are confined to the gut, there is evidence that on occasion the virus may cross the gut and infect other tissues.

Other dsRNA viruses

The virions of most dsRNA viruses have icosahedral symmetry and most of them are naked; an exception is the family *Cystoviridae*, the members of which are enveloped bacteriophages. A problem that dsRNA viruses must overcome is the fact that dsRNA is a potent inducer of a number of cell defence mechanisms including apoptosis, interferon production and RNA silencing. Most of these viruses, including the rotaviruses, have solved this problem by ensuring that the viral dsRNA is always enclosed within virus protein structures, and is therefore never free in the cytoplasm to trigger these defences.

Retroviruses

The retroviruses are RNA viruses that copy their genomes into DNA during their replication. Until the discovery of these viruses it had been dogma that the transfer of genetic information always occurs in the direction of DNA to RNA, so finding that some viruses carry out 'transcription backwards' (reverse transcription) caused something of a revolution. We now know that reverse transcription is carried out, not only by these RNA viruses, but also by some DNA viruses and by uninfected cells. The aim of the current chapter is to provide a general introduction to the retroviruses, which have been found in all classes of vertebrate animal, including fish, amphibians, birds and mammals. The human immunodeficiency viruses (HIV-1 and HIV-2) are retroviruses and the next chapter is devoted entirely to these viruses. Many retroviruses can cause cancer in their hosts.

Retrovirus Virion

The virion contains two copies of the RNA genome, hence the virion can be described as diploid. The two molecules are present as a dimer, formed by base pairing between complementary sequences. The regions of interaction between the two RNA molecules have been described as a 'kissing-loop complex'.

As well as the virus RNA, the virion also contains molecules of host cell RNA that were packaged during assembly. This host RNA includes a molecule of transfer RNA (tRNA) bound to each copy of the virus RNA through base pairing. The sequence in the virus RNA that binds a tRNA is known as the primer binding site (PBS). Each retrovirus binds a specific tRNA. A number of protein species are associated with the RNA. The most abundant protein is the nucleocapsid (NC) protein, which coats the RNA, while other proteins, present in much smaller amounts, have enzyme activities.

Encasing the RNA and its associated proteins is the capsid, which appears to be constructed from a lattice of capsid (CA) protein. The shape of the capsid is spherical, cylindrical or conical depending on the virus. A layer of matrix (MA) protein lies between the capsid and the envelope. Associated with the envelope are two proteins: a transmembrane (TM) protein bound to a heavily glycosylated surface (SU) protein. In most retroviruses the bonds between the TM and SU proteins are non-covalent.

The genes encoding the virus proteins are organized in three major regions of the genome:

1. *gag* (group-specific antigen) – internal structural proteins
2. *pol* (polymerase) – enzymes
3. *env* (envelope) – envelope proteins.

Retrovirus Replication

Attachment and entry

The virion binds to cell receptors via the virus attachment site, which is located on the SU protein. This interaction causes a conformational change in the TM protein that allows a hydrophobic fusion sequence to fuse the virion membrane and a cell membrane. Most retroviruses fuse their membrane with the plasma membrane of the cell, though some are endocytosed and fuse their membrane with an endosome membrane. The structure that is released into the cytoplasm loses some proteins and a reverse transcription complex is formed.

Reverse transcription

Reverse transcription takes place within the reverse transcription complex. Synthesis of both the (-) DNA and the (+) DNA begins at the 3'-OH of a primer RNA. The primer for synthesis of the (-) DNA is the tRNA bound to the genome, while the primer for synthesis of the (+) DNA is a polypurine tract (PPT) in the virus genome. The latter becomes accessible as a result of hydrolysis of the genome RNA from the 3' end by the RNase H, which is an enzyme that specifically digests RNA in RNA–DNA duplexes.

During synthesis of the two DNA strands, each detaches from its template and re-attaches at the other end of the template through base pairing. The DNA that results from reverse transcription (the provirus) is longer than the RNA genome. Each of the termini has the sequence U3–R–U5, known as a long terminal repeat (LTR), one terminus having acquired a U3 sequence and the other a U5 sequence.

Integration of the provirus

The provirus, still associated with some virion protein, is transported to the nucleus as a pre-integration complex. For most retroviruses this can occur only if the cell goes into mitosis, and it is likely that mitosis-induced breakdown of the nuclear membranes is necessary for the pre-integration complex to enter the nucleus. This means that there can be a productive infection only in dividing cells. HIV and related viruses, however, can productively infect resting cells, as the pre-integration complexes of these viruses are able to enter intact nuclei.

One of the virus proteins still associated with the provirus is the integrase; this enzyme cuts the DNA of a cell chromosome and seals the provirus into the gap. The integrated provirus genes may be expressed immediately, or there may be little or no expression of viral genes, in which case a latent infection has been initiated. If a latently infected cell divides, the provirus is copied along with the cell genome and each of the daughter cells has a copy of the provirus.

Transcription and genome replication

The two LTRs of the provirus have identical sequences, but are functionally different; transcription is initiated in one and terminated in the other. Transcription factors bind to a promoter in the upstream LTR, then the cell RNA polymerase II starts transcription at the U3 –R junction. Transcription continues into the downstream LTR. There is a polyadenylation signal in the R region and transcription terminates at the R–U5 junction. Each transcript is capped and polyadenylated. Some transcripts will function as mRNA and a proportion of these become spliced; others will become the genomes of progeny virions.

Translation and post-translational modifications

The *env* gene is translated from spliced mRNAs in the rough endoplasmic reticulum, where glycosylation commences. The Env protein molecules are transported to the Golgi complex, where they are cleaved by a host protease into SU and TM molecules. The two cleavage products remain in close association, and after further glycosylation they are transported to the plasma membrane. The proteins encoded by the *gag* and *pol* genes are translated from genome-length mRNAs into Gag and Gag–Pol polyproteins. Retroviruses require much greater quantities of the Gag proteins than of the Pol proteins, and have evolved mechanisms to synthesize the required amount of each. These mechanisms involve approximately 95 per cent of ribosomes terminating translation after the synthesis of Gag, while the other ribosomes continue translation to synthesize Gag–Pol.

One mechanism, used by murine leukaemia virus, involves reading through a stop codon at the end of *gag*. A 'suppressor tRNA' incorporates an amino acid at the stop codon and translation continues into *pol*, which is in the same reading frame as *gag*. The majority of retroviruses, however, ensure the correct proportions of the Gag and Pol proteins by a ribosomal frameshifting mechanism. Here *gag* and *pol* are in different reading frames and there is a − 1 shift in reading frame before the *gag–pol* junction in about five per cent of translations. This mechanism is used by HIV-1. The Gag and Gag–Pol proteins of most retroviruses are myristylated at their N termini.

Assembly and release of virions

Some retroviruses form immature particles in the cytoplasm that are then transported to the plasma membrane, but most retroviruses assemble their components on the inner surface of the plasma membrane.

The N termini of the Gag and Gag–Pol proteins become anchored to the plasma membrane by the myristyl groups, and the association is stabilized through electrostatic interactions between positive charges in the MA domains and negatively charged phosphate groups in the membrane. The MA domains also bind to the cytoplasmic tails of TM proteins in the membrane.

The NC domains of Gag and Gag–Pol bind the polyproteins to the virus RNA and mediate the formation of the genome dimer. The proteins bind first to a packaging signal near the 5' end of each RNA molecule, and a tRNA binds to the PBS. The RNA then becomes coated with many copies of Gag and a few copies of Gag–Pol.

The immature virion acquires its envelope by budding from the cell surface. At this stage multiple copies of Gag and Gag–Pol are arranged radially with their N termini facing outward and their C termini inward. The late (L) domains of Gag bind host cell factors that are involved in the budding process. During and/or after the budding process the Gag and Gag–Pol polyproteins are cleaved by the virus protease. The cleavage products of Gag form the matrix, the capsid and the protein component of the nucleocapsid, while the Pol cleavage products are the virion enzymes.

Examples of retroviruses

Retroviruses can be classed as either simple or complex, depending on the complexity of their genomes. The simple retroviruses are those that have only the three standard retrovirus genes (*gag*, *pol*, *env*), or in some cases one additional gene, called an oncogene because its expression might result in its host cell developing into a tumour cell. An example of an oncogene is *src* in the genome of Rous sarcoma virus, which infects chickens.

The complex retroviruses have additional genes, the products of which have a variety of functions in the replication cycle. The human immunodeficiency viruses are complex retroviruses; because of their importance the next chapter is devoted entirely to them.

As the names of the viruses imply, many of them of them are causative agents of disease in mammals (feline leukaemia virus), birds (Rous sarcoma virus) and fish (walleye dermal sarcoma virus).

Retroviruses as Gene Vectors

Some genetically modified retroviruses are used as gene vectors. They can introduce genes into a variety of cell types, where the genes are expressed at high levels after integration into the cell genomes. Technologies have been developed for expression of genes in cell cultures and for clinical treatments of genetic disorders and cancers. One of the most commonly used viruses in these applications is murine leukaemia virus. Lentiviral vectors have also been developed; these have the advantage that they can deliver genes into non-dividing cells and tissues.

Patients with the genetic disorder X-linked severe combined immunodeficiency (X-SCID) have been successfully treated with retroviral vectors. The procedure involves taking stem cells from the patient and infecting them with a recombinant retrovirus, the genome of which contains a good copy of the gene that is defective. If the vector provirus is integrated into the cell genome, and if the good copy of the gene is expressed, then the patient is able to develop a normal immune system. The successes, however, have been tempered by the development of cancer in a few treated patients.

Endogenous Retroviruses

It has been known for some time that the genomes of vertebrate animals contain retroviral sequences. The genomes of most of these endogenous retroviruses (ERVs) are defective. Sequencing the human genome has revealed the presence of almost 100 000 human ERV (HERV) sequences, and ERVs have been found in the genomes of other species as they have been sequenced.

Some ERVs are closely related to normal retroviruses (exogenous retroviruses); for example, there are ERVs in mice that have very similar sequences to the genome of mouse mammary tumour virus. It is highly likely that ERVs originated as a result of exogenous retroviruses infecting germ line cells (sperm and/or egg). If one of these cells with an integrated provirus survived to be involved in the reproductive process, then each cell in the body of the offspring would contain a copy of the provirus. Over time ERVs have copied themselves to other locations in the genome, giving rise to families of related ERV elements.

As stated above, most ERVs are defective, so they do not normally replicate. There are circumstances, however, when some ERVs can replicate. Missing functions may be supplied by another ERV or an exogenous retrovirus. Some ERVs do not replicate in cells of the species in which they occur, but are able to replicate in the cells of other species; e.g., some mouse ERVs and some pig ERVs can replicate in human cells. There are also some ERVs that are not defective; they have an intact genome (*gag*, *pol* and *env* genes) and can initiate a productive infection. Because of these findings concern has been expressed that there may be a risk of transmitting retroviruses from pigs into humans if pigs are used as sources of cells, tissues and organs because of shortages of their human counterparts for transplant purposes.

PICORNAVIRUSES

Members of the family *Picornaviridae* are found in mammals and birds. Poliovirus was one of the first viruses to be propagated in cell culture and was also one of the first to be plaque purified. Most picornaviruses grow readily in cell culture, are easy to purify and are stable, making them popular viruses for laboratory studies. The picornaviruses are class IV viruses; their genome is a plus-strand RNA that functions as mRNA once it is released into a host cell. The first virus molecules to be synthesized in an infected cell are therefore proteins, in contrast to all other viruses, where transcription of virus genes must occur before virus protein synthesis can start.

Some Important Picornaviruses

Hepatitis A virus

Hepatitis A is especially prevalent in developing countries with poor sanitation. In most infants and young children infection is asymptomatic or mild, and leads to life-long immunity. When adults become infected about 75% develop jaundice; severe hepatitis is a rare complication, which can be fatal.

Poliovirus

Poliovirus has been the subject of much research effort because of the devastating paralysis that it can cause. In fact, the majority of poliovirus infections are relatively harmless infections of the oropharynx and the gut. Serious disease occurs only after other tissues become infected, resulting in viraemia (virus in the blood) and spread of infection to the central nervous system. This is a very rare event in babies, who still have anti-poliovirus antibodies acquired from their mothers. If, however, there is less poliovirus in the human environment, so that most infections occur after these antibodies have disappeared, then infection of the central nervous system is more likely. Ironically, polio is a disease primarily associated with improving standards of hygiene and sanitation!

Poliovirus infection of the central nervous system may result in meningitis (from which most patients recover completely), encephalitis and/or paralytic poliomyelitis. The latter is due to virus replication in motor neurones of the spinal cord or the brain stem, resulting in paralysis of limbs and/or breathing muscles.

The fear of paralysis was a major motivation for research programmes to develop polio vaccines in the mid-20th century, and these efforts led to the development of inactivated vaccines and live attenuated vaccines. The use of both types of vaccine has proved very effective in preventing polio, so that by the start of the 21st century the annual number of world cases had fallen to 3500, which was 1% of the number of cases in 1988. Polio has been eradicated from many parts of the world and there is hope that the disease will soon be completely eradicated.

Coxsackieviruses

In 1948 investigators in the US town of Coxsackie injected faecal specimens from two suspected polio cases into suckling mice; lesions developed in the skeletal muscles of the mice and the first coxsackieviruses had been isolated. Coxsackieviruses cause a range of medical conditions, including myocarditis (heart disease), meningitis and rashes.

Rhinoviruses

Rhinoviruses are the most common agents causing upper respiratory tract infections in humans. Most children have had at least one rhinovirus infection by the age of 2 years, and in adults rhinoviruses account for about 50% of common colds. Rhinoviruses replicate in the epithelium of the upper respiratory tract, where the temperature is around 33–35°C. For at least some human rhinoviruses the optimum temperature for replication is in this range, though many human rhinoviruses are able to replicate at 37°C and some of these are probably responsible for disease in the lower respiratory tract.

Foot and mouth disease virus

Foot and mouth disease virus has a much wider host range than other picornaviruses; it infects mammals such as cattle, sheep, goats and pigs, causing lesions on the feet and in the mouth.

Outbreaks of foot and mouth disease can have serious economic consequences, as milk yields drop, young infected animals may die and in developing countries animals become unfit for their crucial roles in ploughing and transport. Restrictions on trade in susceptible animal species cause further economic damage. A massive outbreak in the UK in 2001 was brought under control through a series of measures that included the slaughter of more than 6 million animals. The estimated cost of the outbreak to the nation was over 8 billion pounds.

Picornavirus Virion

Picornaviruses are small RNA viruses of relatively simple structure. The RNA is enclosed by a capsid, which is roughly spherical and has a diameter of about 25–30 nm.

Capsid

The picornavirus capsid has icosahedral symmetry and is made from 60 copies each of four virus proteins (VP1–4, numbered from the largest to the smallest). Each of the proteins VP1–3 contains an eight-stranded β-barrel, like many virus capsid proteins, including those of the parvoviruses. Parts of VP1–3 are at the surface of the virion, while the N termini of VP1–3 and all 60 VP4 molecules are completely internal.

In many picornaviruses, including polioviruses and rhinoviruses, there is a deep cleft around each of the 12 vertices of the icosahedron. These clefts, often referred to as canyons, were discovered using X-ray crystallography and cryo-electron microscopy. They are approximately 2 nm deep, which, for a virion of this size, is a significant depth, being approximately seven per cent of the virion diameter. This is relatively much greater than the depth of the Grand Canyon on the surface of the Earth! The canyons, which are lined by the C termini of VP1 and VP3 molecules, contain the virus attachment sites.

Evolution of picornaviruses has generated a lot of variability in the capsid proteins, some of which is reflected in the existence of distinct serotypes. This antigenic variability creates problems for the development of vaccines against these viruses, problems that have been overcome for poliovirus and foot and mouth disease virus, but not for the rhinoviruses.

Genome

The picornavirus genome is composed of a 7–8 kb ssRNA. Covalently linked to the 5' end of the RNA is a small (2–3 kD) protein known as VPg (virus protein, genome linked). The covalent link is via the –OH group of a tyrosine residue at position 3 of VPg. The 3' end of the RNA is polyadenylated.

The genome consists of one very large ORF flanked by untranslated regions. Within the untranslated region at the 5' end there is much secondary structure. For example, in poliovirus RNA there are six domains, five of which form an internal ribosome entry site (IRES).

Picornavirus replication

A model of picornavirus replication has been developed, though uncertainty still surrounds many of the details. Many components of the model stem from studies with poliovirus, which replicates in a much wider range of cell types in culture than it does in the body. Information derived from studies with other picornaviruses will also be mentioned.

Attachment

The cell receptors for a number of picornaviruses have been characterized, including those for the rhinoviruses and poliovirus. The poliovirus receptor is the glycoprotein CD155. Its function is not known, but it is expressed on many celltypes. CD 155 is a member of the immunoglobulin super-family of molecules with three immunoglobulin-like domains; the virus attachment site is located in the outermost domain.

CD 155 is found only in humans and some other primate species. In the early days of poliovirus research none of the common laboratory animals could be infected with the virus, but in the early 1990s transgenic mice expressing CD155 were developed. These animals were found to be susceptible to infection with all three serotypes of poliovirus and they have been used in studies of replication and pathogenesis.

For those picornaviruses like poliovirus with canyons on the virion surface there are virus attachment sites located in pockets at the canyon bases. The canyons are too narrow for access by antibodies so the virus attachment sites are protected from the host's immune surveillance, while the remainder of the virion surface can mutate to avoid the host's immune response. The virus attachment sites of foot and mouth disease virus, however, are not in canyons but are on surface protrusions.

Binding of a virion to receptors results in major changes to the capsid structure: the N termini of VP1 move from the interior to the exterior surface of the capsid and VP4 is lost from the virion. Binding of free CD155 to poliovirus particles also causes loss of VP4.

Entry

Several modes of genome entry into the host cell have been proposed for picornaviruses. One mode involves transfer of the RNA from the virion into the cytoplasm at the plasma membrane, leaving the capsid at the cell surface. Other modes involve endocytosis, followed by either release from the capsid after disruption of the endosome membrane, or release of RNA from the capsid then transport across the endosome membrane. There is evidence that foot and mouth disease virus and some rhinoviruses enter the cell by endocytosis. Once the virus genome is free in the cytoplasm the VPg is removed from the 5' end by a cell enzyme.

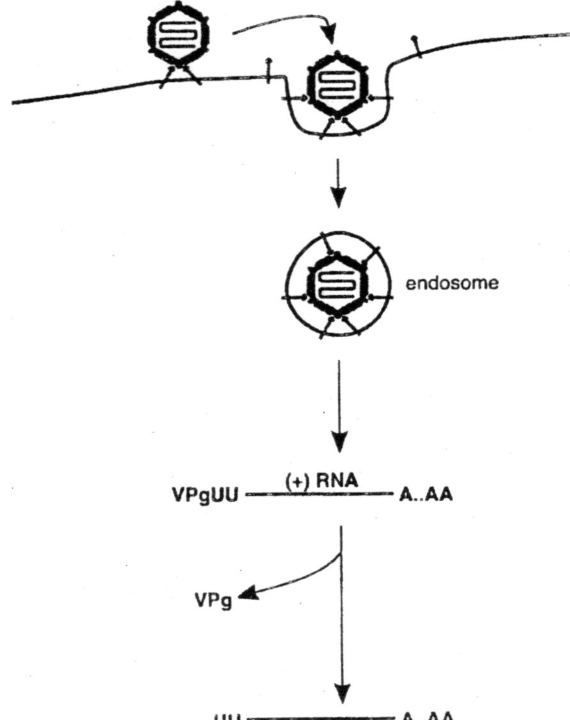

Fig. 14.5. Entry of picornavirus virion by endocytosis.

Translation and post-translational modifications

The virus genome functions as mRNA, but as the RNA is not capped the preliminary stages of translation are different to those for capped mRNAs. The 40S ribosomal subunit binds not at the 5' end of the RNA, but internally at the IRES. Most of the eukaryotic initiation factors (eIFs, normally involved in cap-dependent translation) are involved, but a notable exception is the cap-binding protein eIF4E.

Early investigation found that the number of virus proteins synthesized in the infected cell seemed to be greater than could be encoded by the small genome of a picornavirus. The puzzle was solved when it was found that the genome encodes a single polyprotein, which undergoes a series of cleavages to give rise to all the structural and non-structural proteins.

Virus-coded proteases perform the cleavages, some of which are intramolecular. The polyprotein is first cut to yield P1, P2 and P3. P1 becomes myristylated at the N terminus before being cleaved to VP0, VP3 and VP1, the proteins that will form procapsids; VP0 will later be cleaved to produce VP2 and VP4. Other cleavage products include 3B (VPg), 2C (an ATPase) and 3D (the RNA polymerase). The production of the proteins via a polyprotein precursor might suggest that equimolar amounts result, but not all cleavages take place with equal efficiency. The amounts of the proteins therefore vary; for example, greater amounts of the capsid proteins are produced than of 3D (the RNA polymerase).

Once RNA replication is under way, significant amounts of (+) RNA are produced; some of this RNA functions as mRNA and large quantities of virus proteins are then synthesized. Because the RNA is large (7–8 kb) there may be up to 60 ribosomes translating from each mRNA.

Transcription/genome replication

RNA synthesis occurs in replication complexes that contain cell proteins, as well as virus proteins and RNA. Using immune electron microscopy it has been shown that the replication complexes are associated with the cytoplasmic surfaces of membranous vesicles that develop in the cytoplasm of infected cells. Large numbers of vesicles form, derived mainly from the endoplasmic reticulum. Each vesicle is 200–400 nm in diameter and is bounded by two lipid bilayers. All known class IV viruses of eukaryotes replicate their RNA in similar membrane-associated complexes.

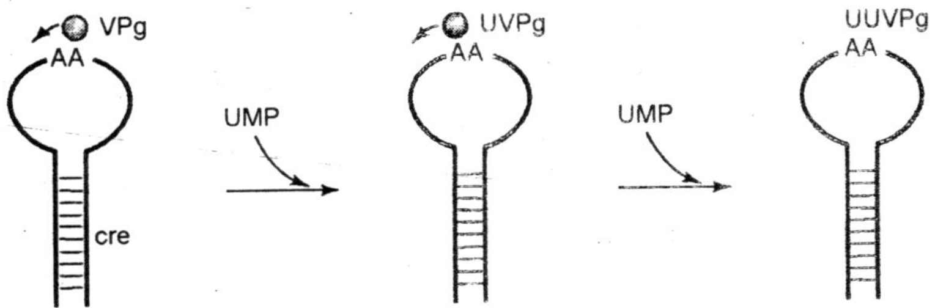

Fig. 14.6. Uridylylation of poliovirus VPg.

In order for the infecting (+) RNA to be replicated, multiple copies of (-) RNA must be transcribed and then used as templates for (+) RNA synthesis. The primer for both (+) and (-) strand synthesis is the small protein VPg, which is uridylylated at the hydroxyl group of a tyrosine residue by the poliovirus RNA polymerase. These reactions take place at a cis-acting replication element (*cre*) located in a stem-loop in the virus genome; an AA sequence within the loop forms the template for UU synthesis. There is evidence that UU-VPg acts as the primer for (-) RNA synthesis, but it is uncertain whether the primer for (+) RNA synthesis is VPg or a uridylylated derivative.

Much of the RNA in the replication complexes is present as replicative intermediates (RIs). An RI is an association of RNA molecules consisting of a template RNA and several growing RNAs of varying length. Thus some RIs consist of a (+) RNA associated with growing (-) RNAs, while other RIs consist of a (-) RNA associated with growing (+) RNAs. Also associated with RIs are copies of the RNA polymerase.

The poly(A) sequence at the 3' end of the plus strand is not synthesized in the usual way. It is transcribed from a poly(U) sequence at the 5' end of the minus strand.

It has been estimated that in a poliovirus-infected cell up to 2500 RNA molecules are synthesized per minute. The virus controls RNA synthesis in such a way that about 50 times more (+) RNA than (-) RNA is made. Some of the (+) RNA molecules are used as templates for further (-) RNA synthesis, some function as mRNA and some are destined to be the genomes of progeny virions.

Assembly and exit

Five copies each of VP0, VP3 and VP1 assemble into a 'pentamer' and 12 pentamers form a procapsid. Each procapsid acquires a copy of the virus genome, with VPg still attached at the 5' end. At around this time the 60 copies of VP0 are cleaved into VP4 and VP2.

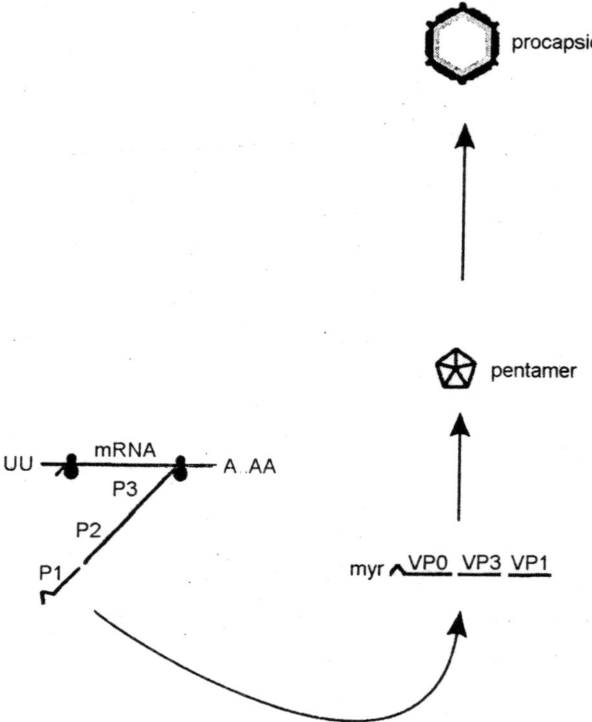

Fig. 14.7. Procapsid assembly.

Infected cell cultures develop cytopathic effects; for example, Vero cells (a monkey cell line) infected with poliovirus become rounded. Lysis of the cell releases the virions; approximately 10^5 virions are produced in each poliovirus-infected cell.

Inhibition of host gene expression

Soon after infection, the expression of host cell genes is inhibited. In poliovirus-infected cells it has been demonstrated that all three host RNA polymerases are rapidly inhibited, thus terminating transcription. Translation from pre-existing cell mRNAs is inhibited as a result of the cleavage of a translation initiation factor by the virus protease 2A. Virus protein synthesis is unaffected as it is initiated by a cap-independent mechanism.

Overview of picornavirus replication

The cell nucleus is not shown, as all stages of virus replication take place in the cytoplasm. It has been demonstrated that poliovirus

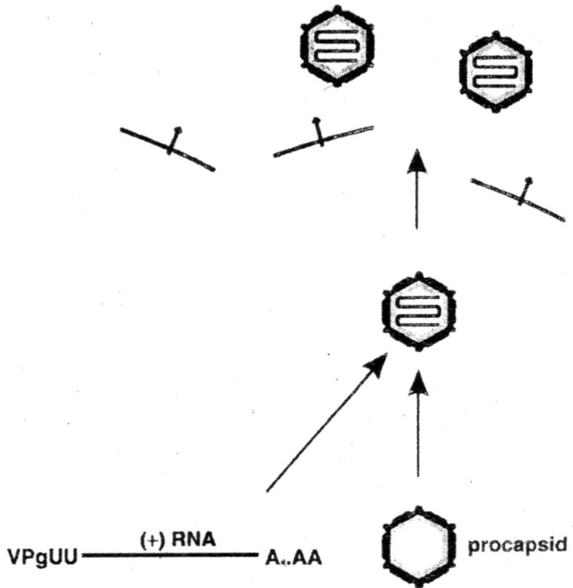

Fig. 14.8. Picornavirus assembly and exit

can replicate in enucleated cells, indicating that the virus has no requirement for the nucleus. It is almost certain, however, that some virus proteins enter the nucleus and are involved in the inhibition of cell transcription.

Picornavirus Recombination

If a cell becomes infected with two strains of a picornavirus, then recombinant viruses may be present amongst the progeny virus from that cell. A recombinant virus is one that has part of its genome derived from one virus and part derived from another virus.

The oral polio vaccine contains attenuated strains of all three poliovirus serotypes. Administration of this vaccine can lead to co-infections of gut cells with two serotypes, and recombinants between serotypes may be formed. If a wild type poliovirus is also present in the gut then recombination may occur between the wild type virus and one of the vaccine strains. There have been a number of documented cases where recombinants have caused polio.

Picornavirus Experimental Systems

This section introduces some techniques that have been developed during work with picornaviruses. Reverse genetics techniques that permit manipulation of RNA genomes are discussed first, followed by techniques that enable the synthesis of infectious virus particles in a cell-free system.

Reverse Genetics

Techniques that have been developed for manipulating DNA genomes can be applied to RNA genomes if the RNA is reverse transcribed to DNA. A defined mutation can be introduced into a gene in the DNA to investigate the function of the gene. After mutation the DNA is transcribed back to RNA, which is then transfected into cells in culture. Virus replication in the transfected cells leads to the production of mutant virus.

In one application of this technique with poliovirus, it was demonstrated that myristylation of VP0 is essential for virion assembly to proceed. A genome was produced in which the glycine residue at the N terminus of VP0 was mutated to alanine; VP0 of the mutant virus could not be myristylated and virion assembly was inhibited.

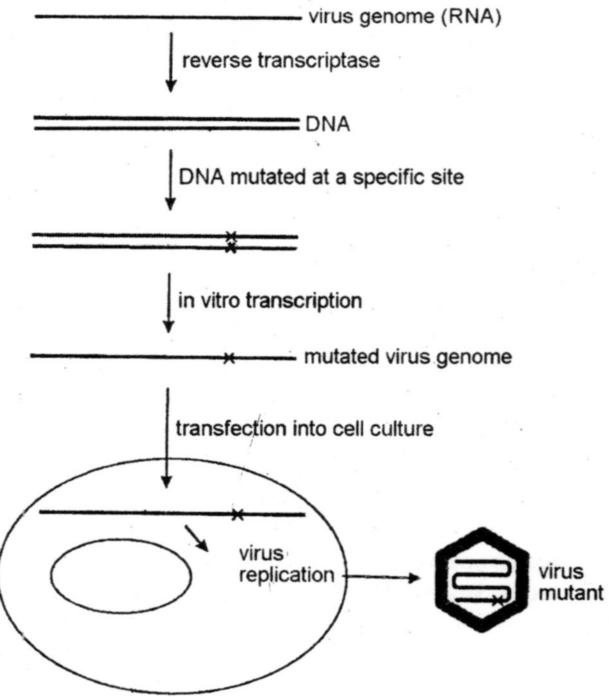

Fig. 14.9. Production of a picornavirus mutant by reverse genetics.

Cell-free synthesis of infectious virus

In 1991 a group of workers recovered infectious poliovirus from a HeLa cell extract to which they had added purified poliovirus RNA. This was the first report of the production of infectious virus outside cells. Cell-free systems based on cytoplasmic extracts have subsequently been used to investigate features of the poliovirus replication cycle that are impossible to dissect using cell-based systems. In 2002 it was reported that infectious poliovirus had been produced in a process that started with the synthesis of a copy DNA (cDNA) of the virus genome.

The process started with the synthesis of discrete sequences of DNA, which together spanned the sequence of a poliovirus genome, then the sequences were ligated to form a cDNA of the virus genome. RNAs (virus genomes) were transcribed from the cDNA and inoculated into a cell extract; infectious poliovirus was subsequently found in the cell extract. This was the first time that infectious virus had been produced from molecular building blocks in the absence of cells.

Other Plus-strand RNA Viruses

The genomes of plus-strand RNA viruses are translated in the infected cell prior to their transcription. Because the ribosomes in eukaryotic cells usually terminate after translation of the first ORF, the viruses need strategies to ensure that all the proteins encoded in the infecting genome are translated.

1. *Polyprotein.* All the genetic information is encoded in one ORF; this is translated to produce a polyprotein which is cleaved to produce the individual virus proteins.

   Two other strategies are used by plus-strand RNA viruses.

1. *Subgenomic mRNAs.* The genome has two or more ORFs. The RNA-dependent RNA polymerase is encoded by the ORF at the 5' end of the genome so that it can be translated from the infecting genome. The other ORF(s) are transcribed into subgenomic mRNA(s) that have the same 3' end as the genome.
2. *Segmented genome.* There is one ORF in each RNA segment.

The only ssRNA phages that are known are classified in the family *Leviviridae*. Viruses in this family, such as phage MS2, have genomes with characteristics of bacterial mRNA; the genomes are polycistronic, with all of the ORFs translated by internal initiation.

RHABDOVIRUSES

The rhabdoviruses have minus-strand RNA genomes in the size range 11–15 kb. The name of these viruses is derived from the Greek word *rhabdos*, which means a rod. The virions of some rhabdoviruses, especially those infecting plants, are in the shape of rods with rounded ends, while others, especially those infecting animals, are bullet shaped.

Rhabdoviruses are found in a wide range of hosts, including mammals, fish, plants and insects, and many rhabdoviruses are important pathogens of animals and plants. The rhabdoviruses constitute the family *Rhabdoviridae*, which contains a number of genera.

Many rhabdoviruses have very wide host ranges and replicate in the cells of diverse types of host, especially the so-called 'plant' rhabdoviruses, which replicate in their insect vectors as well as in their plant hosts.

Before looking at the structure and replication of rhabdoviruses, we consider two important rhabdoviruses, rabies virus and vesicular stomatitis virus (VSV).

Some Important Rhabdoviruses

Rabies virus

Rabies virus, like many rhabdoviruses, has an exceptionally wide host range. In the wild it has been found infecting many mammalian species, while in the laboratory it has been found that birds can be infected, as well as cell cultures from mammals, birds, reptiles and insects. Infection with rabies virus normally occurs as a result of virus in saliva gaining access to neurones through damaged skin. The infection spreads to other neurones in the central nervous system, then to cells in the salivary glands, where infectious virus is shed into the saliva. Each year rabies kills large numbers of humans, dogs, cattle and other animal species; precise numbers are not known, but for humans it is estimated that rabies causes about 60000 deaths annually. Most rabies infections of humans are acquired via bites from rabid dogs, though a few people have become infected after receiving an organ transplant from a rabies-infected individual.

Rabies is endemic in wild animals in many parts of the world. In many regions a single animal species serves as the major reservoir; in Western Europe the major reservoir is the red fox.

Vaccines have been developed to provide protection to humans (e.g. veterinary surgeons), domestic animals (especially dogs) and wild animals (e.g. foxes) at risk from rabies virus infection. Rabies vaccines have been incorporated into food baits attractive to wild mammals, and dropped from aircraft over fox-inhabited regions in Europe and over coyote- and raccoon-inhabited regions in the US. The first vaccine

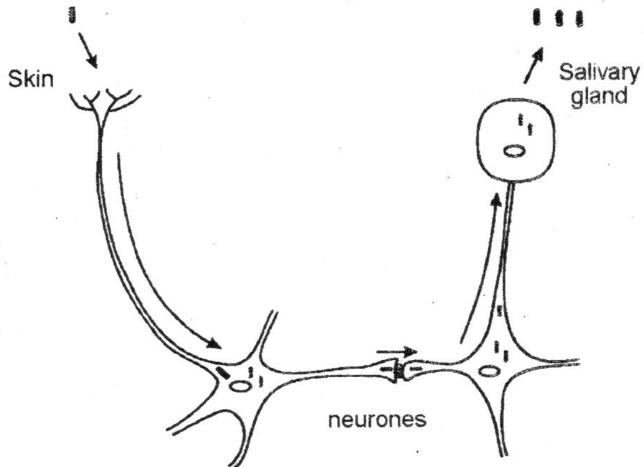

Fig. 14.10. Rabies virus infection of the animal body.

to be used was an attenuated vaccine, but more recent vaccines have contained a recombinant vaccinia virus that expresses the rabies virus G protein. Vaccination of wild mammals has been very successful in bringing rabies under control in a number of countries.

Rabies is normally absent from the UK. In the past, this status was maintained through the requirement for a quarantine period for certain animal species, including dogs, on entry to the country. That policy has been largely replaced with a 'pet passport scheme', which involves giving rabies vaccine to animals prior to entry, and implanting an identifying microchip in each vaccinated animal.

Many viruses related to rabies virus have been found in bats around the world, and have been classified in the genus *Lyssavirus* along with the original rabies strains. There are occasional cases of human rabies resulting from bites from infected bats. One such victim was David McRae, a licensed bat handler in Scotland, who died in 2002 after being bitten by an insectivorous bat.

Vesicular stomatitis virus

VSV causes disease in a variety of animals, including cattle, horses, sheep and pigs, affected animals developing lesions on the feet and in the mouth similar to those in foot and mouth disease. The disease can result in significant economic damage due to decreased milk and meat production, and the imposition of quarantines and trade barriers. Vesicular stomatitis is endemic in the tropics and there are cyclic epidemics in some temperate areas, but it has never been found in the UK.

VSV has a very wide host range. As well as infecting domestic livestock there is evidence of infection in wild animals including bats, deer and monkeys. This evidence is the presence in these animals of neutralizing antibodies to the virus. VSV has been isolated from a number of insect species, including mosquitoes, sand flies and black flies. Its natural cycle is unknown, but it is possible that it is transmitted between mammals by one or more of these types of insect.

In the laboratory VSV can replicate in cell cultures derived from mammals, birds, fish, insects and nematode worms. Much of our understanding of rhabdovirus structure and replication comes from studies with VSV, which is much safer than rabies virus to work with. Three species of VSV are recognized.

Rhabdovirus Virion and Genome Organization

The rhabdovirus virion is an enveloped, rod- or bullet-shaped structure containing five protein species.

The nucleoprotein (N) coats the RNA at the rate of one monomer of protein to nine nucleotides, forming a nucleocapsid with helical symmetry. Associated with the nucleocapsid are copies of P (phosphoprotein) and L (large) protein. The L protein is well named, its gene taking up about half of the genome. Its large size is justified by the fact that it is a multifunctional protein, as will be described later. The M (matrix) protein forms a layer between the nucleocapsid and the envelope, and trimers of G (glycoprotein) form spikes that protrude from the envelope.

The genomes of all rhabdoviruses encode these five proteins. Many rhabdoviruses encode one or more proteins in addition to these.

Rhabdovirus Replication

Attachment and entry

A rhabdovirus virion attaches to receptors at the cell surface and is then taken into the cell by clathrin-mediated endocytosis. The G protein spikes are involved both in the attachment to cell receptors and in the membrane fusion. The nucleocapsid is released into the cytoplasm after the membranes of the virion and the endosome have fused.

Transcription

Once the RNA and its associated proteins (N, P and L) are free in the cytoplasm transcription of the virus genome can begin. A plus-strand leader RNA, the function of which is uncertain, and five mRNAs are synthesized.

Transcription is carried out by an RNA-dependent RNA polymerase activity that resides, along with four other enzyme activities, in the L protein. The polymerase is active only when L is complexed with P protein in the ratio 1L:3P. The requirement for P was demonstrated in experiments with VSV in which P and L were purified. The individual purified proteins were found to be lacking in polymerase activity, which was restored when the two proteins were mixed.

Also associated with the L protein are enzyme activities that cap and polyadenylate the mRNAs. The virus supplies these activities, as the cell enzymes are present only in the nucleus. Capping and polyadenylation of the virus mRNAs proceed by mechanisms different to those carried out by the cell, though the end results are the same: each mRNA is capped at the 5' end and polyadenylated at the 3' end.

As the L–P complex moves along the template RNA from the 3' end to the 5' end a newly synthesized RNA molecule is released at each intergenic sequence. The first RNA synthesized is the leader RNA and the remainder are mRNAs. Before release, each mRNA is polyadenylated by the poly(A) polymerase activity of L. It is thought that the L–P complex 'stutters' at the 5′ end of each gene, where the sequence UUUUUUU is transcribed as about 150 adenylates. The enzyme resumes transcription when it recognizes the start of the next gene.

The virus does not need equal amounts of all the gene products; for example, it needs many copies of N protein to coat new RNA, but relatively few copies of L protein. It controls the expression of its genes by controlling the relative quantities of transcripts synthesized. As the enzyme complexes move along the (–) RNA approximately 30% of them detach at the end of each gene, so that fewer and fewer enzyme complexes remain associated with the template as they progress towards the 5' end. Thus, many copies of N are transcribed, but relatively few copies of L.

Translation

The virus proteins are translated on free ribosomes, except for the G protein, which is translated in the rough endoplasmic reticulum.

As their names imply, the phosphoprotein (P) and the glycoprotein (G) undergo post-translational modification. One-sixth of the residues in VSV P protein are serine and threonine, and many of these are phosphorylated. The phosphorylation takes place in two steps, the first performed by a cell kinase and the second by the kinase activity of the L protein. After phosphorylation, trimers of P are formed.

Glycosylation of G protein commences in the rough endoplasmic reticulum, where core monosaccharides are added, and is completed in the Golgi complex.

Genome replication and secondary transcription

The minus-strand virus genome is replicated via the synthesis of complementary (+) RNA molecules, which then act as templates for the synthesis of new copies of (-) RNA. Replicative intermediates can be detected in infected cells, as with the plus-strand RNA viruses. The initiation of RNA synthesis does not require a primer.

We noted earlier that the leader RNA and the individual mRNAs are produced as a result of the RNA polymerase recognizing a termination signal at each intergenic sequence of the template and at the end of the L gene. During genome replication, however, the enzyme must remain associated with the template to produce genome-length (+) RNA. Another difference between the two processes is that during genome replication the newly synthesized (+) RNA quickly becomes coated with N protein, whereas the mRNAs are not coated. The genome and the genome-length (+) RNA are never present in the cell as naked molecules, but are always associated with N protein, which protects them from ribonucleases. This is true for all minus-strand RNA viruses.

The differences between the processes that result in synthesis of mRNAs and genome-length (+) RNA from the same template are not understood. One suggestion is that there may be differences in the components of the enzyme complexes involved in the two processes, one complex acting as a 'transcriptase' and the other acting as a 'replicase'.

A rhabdovirus-infected cell synthesizes about 4 to 10 times more copies of (-) RNA than genome-length (+) RNA. Some copies of the (-) RNA are used as templates for further transcription (secondary transcription) so that the amounts of virus gene products in the cell can be boosted, while some become the genomes of progeny virions.

Assembly of virions and exit from the cell

It was noted above that both minus strands and plus strands of genome-length RNA are coated with N protein. Only coated minus strands, however, are selected to form virions, because of the presence of a packaging signal at the 5' end of the minus strand.

The M protein plays several important roles in the assembly process. It condenses the nucleocapsid into a tightly coiled helix and it links the nucleocapsid with a region of the plasma membrane into which copies of the G protein have been inserted. Virions bud from these regions of the plasma membrane, acquiring their envelopes in the process. The M protein has a late (L) domain that binds cell proteins involved in the budding process.

Inhibition of host gene expression

Rhabdovirus infection of a cell results in strong inhibition of host gene expression. The M protein, whose important roles in virion assembly have just been described, appears to play major roles in this inhibition. There is evidence that the M protein inhibits transcription by all three host RNA polymerases and that it blocks intracellular transport of cell RNAs and proteins. One effect of these activities in animal cells is that the synthesis of interferon is inhibited.

Overview of rhabdovirus replication

There appears to be no significant role for the nucleus in the replication of most rhabdoviruses. Enucleated cells support VSV replication, with only a small drop in yield compared with normal cells. This is not true for all rhabdoviruses, however, as members of the genus *Nucleorhabdovirus* replicate in the nucleus of their plant and insect hosts, with virions budding through the inner nuclear membrane.

Other Minus-strand RNA Viruses

Most viruses with minus-strand RNA genomes have animal hosts, while a few have plant hosts. Amongst them are three of the world's major human pathogens: influenza, measles and respiratory syncytial viruses. These three viruses cause millions of cases of serious disease and approximately a million deaths each year.

Measles virus is one of the main causes of childhood death in developing countries and is still responsible for some deaths in industrialized countries. Infection results in immunosuppression, which renders the host more susceptible to secondary infections with a range of bacterial and viral pathogens, and these cause most measles-associated deaths. Measles virus can also cause lethal infections of the central nervous system.

Viruses in the families *Paramyxoviridae* and *Filoviridae*, like those in the *Rhabdoviridae*, have non-segmented genomes that are transcribed to mRNAs by their RNA-dependent RNA polymerases terminating and reinitiating at intergenic sequences. These families form part of the order *Mononegavirales*, which are viruses with monopartite genomes composed of negative sense RNA. Viruses in the families

Orthomyxoviridae and *Bunyaviridae*, on the other hand, have segmented genomes.

Viruses with Ambisense Genomes

The RNA segments of some of the segmented genome viruses are ambisense, where one or more of the RNA segments each encodes two genes, one in the plus sense and one in the minus sense. This is the case for the two genome segments of viruses in the family *Arenaviridae*. It is also the case for some of the three genome segments of some members of the family *Bunyaviridae*, such as tomato spotted wilt virus; two of the three genome segments of this virus are ambisense. The minus-sense gene of an ambisense RNA is expressed by transcription of a mRNA. The plus-sense gene is expressed by synthesis of an RNA complementary to the genome, followed by transcription of the mRNA for that gene.

Reverse Genetics

Reverse genetics have been used to investigate rhabdoviruses and other minus-strand RNA viruses, though the procedures are not as straightforward as with the plus-strand RNA viruses. Minus-strand RNA is not infectious, so techniques had to be devised that will not only generate virus genomes from cDNA, but also supply the RNA polymerase and the nucleoprotein that coats the newly synthesized RNA. After much painstaking work, there are now procedures that enable the recovery of infectious minus-strand RNA virus from a cDNA.

As with the plus-strand RNA viruses, a defined mutation can be introduced into a gene in the cDNA to investigate its function in the replication cycle and its role, if any, in virus virulence. Reverse genetics are also being explored as a tool to engineer virus strains with reduced virulence for use in vaccines.

Hepadnaviruses

The hepadnaviruses got their name because they cause hepatitis and they have *DNA* genomes. They are known as hepatitis B viruses (HBVs) and are classified in the family *Hepadnaviridae*. Some members infect mammals and some infect birds; examples include woodchuck HBV and heron HBV. The best-known hepadnavirus is that which infects humans; it is commonly referred to as HBV, and is of major importance as an agent of disease and death. Duck HBV, on the other hand, is non-pathogenic in its natural host.

The hepadnaviruses are especially fascinating for two reasons. First they have very small genomes, which are used with great economy to encode the virus proteins and to control expression of the virus genes. Second, their DNA genomes are replicated via an RNA intermediate. In other words, their replication involves reverse transcription, so they are very different from DNA viruses that replicate their DNA directly to DNA. The discovery of the mode of replication of the hepadnavirus genome led to the creation of Baltimore class VII.

DNA viruses that replicate via RNA have also been found in plants; an example is cauliflower mosaic virus. These viruses, together with the hepadnaviruses, have been termed pararetroviruses. This chapter will concentrate on HBV.

Importance of HBV

No-one knows how many people are infected with HBV, but the figure is probably around 400 million. The majority are in Asia and many are in Africa. There are high percentages of people infected in the far north of North America and Greenland, but because of the low populations of these regions the numbers infected are relatively low.

Virus is present in the blood and semen of infected individuals and the modes of transmission generally parallel those for HIV transmission. There are over 50 million new HBV infections each year, the majority in babies who acquire the infection from their mothers. Over 8 million infections per year are thought to result from the re-use of syringes and needles for injections, mainly in the developing world.

Many HBV infections result in mild symptoms or are asymptomatic, especially in children. It is in children, however, that HBV infection is most likely to become persistent, with 90–95 per cent of those infected as newborn infants becoming longterm carriers, compared with 1–10 per cent of those becoming infected as adults.

Individuals who are persistently infected with HBV may remain healthy for much of the time, but some develop severe hepatitis, which may lead to cirrhosis and eventually to liver cancer. These diseases resulting from HBV infection cause about one million deaths per year.

HBV Virion

The virion is roughly spherical, with a diameter of about 42 nm. The main virion components are an envelope enclosing a capsid, inside which is the DNA and P (polymerase) protein.

DNA

The genome is made up of two strands of DNA, one of which is incomplete; hence the DNA is partly single- stranded and partly double-stranded. A short sequence is triple-stranded as a result of a complementary sequence at the 5' ends, and this results in the DNA having a circular conformation. The genome is very small, with a length of about 3.2 kb(p). At the 5' end of each of the DNA strands there is a covalently linked molecule: a capped RNA on the short strand and a protein (P) on the long strand.

Polymerase (P) protein

The virion contains at least one complete molecule of P (polymerase) protein. The N terminus of P constitutes a 'terminal protein' domain; it is not known whether the protein linked to the DNA is a complete molecule of P or only the terminal protein. The terminal protein domain is separated by a 'spacer' from a domain with reverse transcriptase activity. At the C terminus there is a domain with ribonuclease H (RNase H) activity. P also has DNA-dependent DNA polymerase activity.

Capsid

The capsid, which has icosahedral symmetry, has holes in it and short spikes protrude from its surface. It is constructed from dimers of the C (core) protein, which is largely α-helical, unlike the capsid proteins of many other viruses. The C terminus of the C protein is highly basic due to the presence of a large number of arginine residues; this region binds the virus genome.

Envelope

The virion envelope contains three protein species designated as small (S), medium (M) and large (L). The M and L proteins are longer versions of the S protein, which is the most abundant of the three. The surface regions of the envelope proteins constitute an antigen known as hepatitis B surface antigen (HBsAg). Each envelope protein has one or more glycosylation sites, though not all molecules are glycosylated. The virus attachment site is near the N terminus of the L protein, but only about 50 per cent of the L molecules have the N terminus on the outside of the virion; the N termini of the remaining L molecules are on the inside bound to the capsid.

Non-infectious Particles

An unusual and intriguing feature of HBV infection is the presence in the blood of not only virions, but also large quantities of non-infectious particles that have been released from infected liver cells. These particles are composed of lipid and virus envelope proteins, but they do not contain nucleocapsids. Some of the particles are spheres

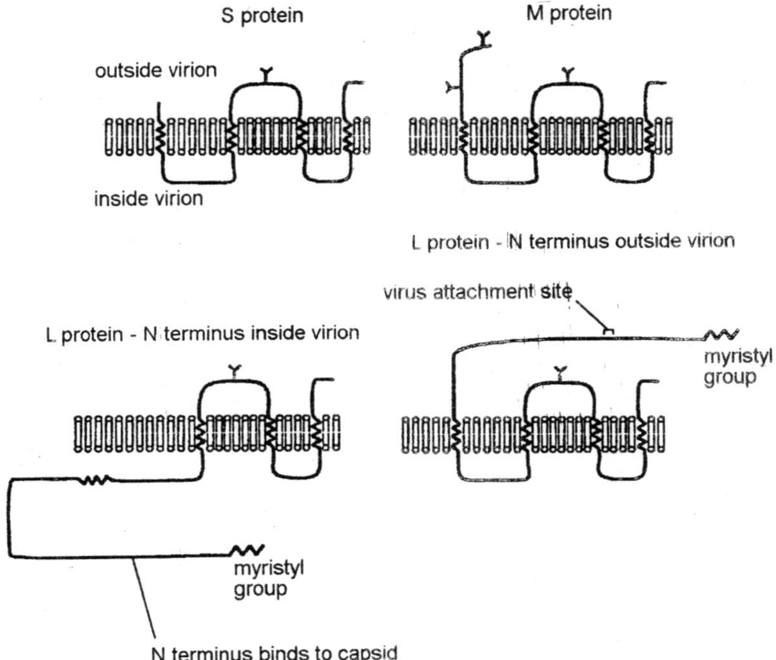

Fig. 14.11. Arrangement of HBV small (S), medium (M) and large (L) proteins in the virion envelope.

and some are filaments; both have diameters of 22 nm and the filaments have variable lengths up to 200 nm.

All the particles (virions and non-infectious particles) are much more abundant in the blood than in the liver, and the non-infectious particles, especially the spheres, vastly outnumber the virions. Presumably the virus has a good reason for inducing the production of such large amounts of non-infectious material, but that reason is not understood. It has been suggested that it might be a decoy for antibodies, thereby providing virions with some protection from the host's immune system.

Soluble Virus Protein

In addition to the virions and non-infectious particles a soluble virus protein is found in the blood of some infected individuals. This protein is known as hepatitis Be antigen (HBeAg). It is similar to the C protein, but has 10 extra amino acid residues at the N terminus and lacks 34 amino acid residues at the C terminus. The function of HBeAg, like that of the small spheres and filaments, is not known.

HBV Genome

At 3.2 kb, the HBV genome is very small, though there are viruses with smaller genomes. There are four ORFs, from which seven proteins are translated, so a large amount of coding information is packed into the small genome. The virus cleverly achieves this by using every nucleotide in the genome for protein coding and by reading more than half of the genome in two reading frames. The P ORF, which occupies about 80 per cent of the genome, overlaps the C and X ORFs and the entire S ORF is within the P ORF. A further way in which the virus maximizes its coding capacity is by expressing the L protein in two different conformations that have different functions. In one conformation L protein molecules act as virus attachment proteins, while in the other they bind the virion envelope to the capsid.

As the entire genome is involved in coding for protein, it follows that all the regulatory sequences, such as promoters, are within protein coding sequences. The genome contains direct repeats of 11 nucleotides known as DR1 and DR2.

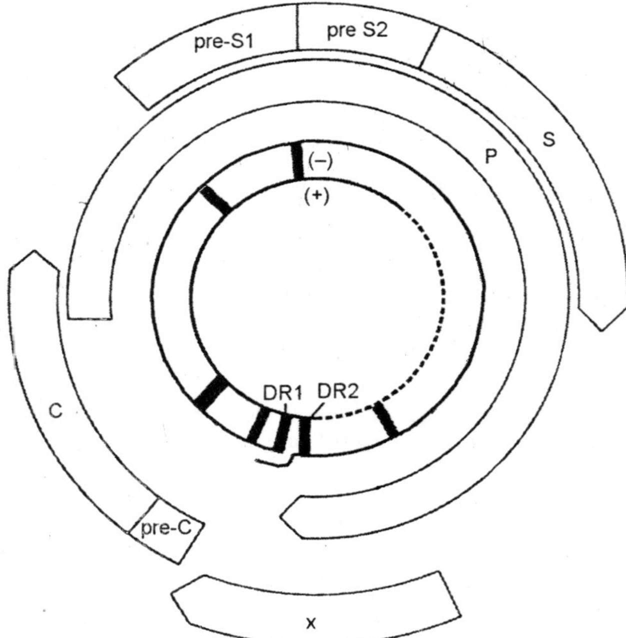

Fig. 14.12. HBV genome organization. The incomplete (+) strand and the complete (-) strand of the DNA are shown surrounded by the four ORFs.

Expression of the pre-S 1-pre-S2-S region gives rise to three proteins. Translation of the S region produces the S protein, translation of pre-S2-S produces the M protein and translation of the complete ORF produces the L protein. Similarly, expression of the pre-C-C region gives rise to two proteins. Even though the virus maximizes the use of its small genome it encodes only seven proteins, so it is heavily dependent on host cell functions. Host proteins that have been demonstrated to play roles in the replication of HBV include enzymes, transcription factors and chaperones.

HBV Genetic Groups

Sequencing of many HBV isolates has revealed eight genetic groups (genotypes A-H) and these human viruses are related to similar viruses in other primate species. The genetic groups are fairly restricted geographically, e.g. genotype A predominates in Northern Europe while genotypes B and C are in Asia.

HBV Replication

Hepatocytes (liver cells) are the host cells for HBV in the body. In the laboratory, primary cell cultures of human hepatocytes support replication, but unfortunately none of the established cell lines derived from liver tumours can be infected by HBV virions. Some cell lines, however, can be infected using HBV DNA (a procedure known as transfection), though this does not permit study of the processes of entry and uncoating. Some hepadnaviruses, such as the woodchuck and duck HBVs, replicate in cell lines. Much of our knowledge of hepadnavirus replication is therefore derived from studies with cell cultures, either transfected with human HBV DNA or infected with animal hepadnaviruses.

Attachment

The identity of the hepatocyte receptor to which HBV attaches is not yet known, though various candidates have been proposed, such as polymeric IgA and annexin V. The virus attachment site is on the L protein.

Entry

The virion is endocytosed then the nucleocapsid is released from the endosome by fusion of the virion and endosome membranes. The nucleocapsid is shown entering the nucleus, as evidence has been published that HBV capsids can pass through nuclear pores. It is not certain, however, whether this happens, or whether uncoating occurs within a nuclear pore, or whether the genome is released into the nucleus with the capsid remaining in the cytoplasm. Once the virus genome is free in the nucleus it is converted into a circular DNA molecule. The covalently bound P protein is removed from the 5' end of the minus strand, which is shortened to remove the third strand of the triple-stranded region. The RNA is removed from the 5' end of the plus strand, while DNA synthesis at the 3' end makes the entire molecule double-stranded. The ends of each strand are then ligated to form covalently closed circular DNA (cccDNA). It is uncertain whether all these modifications to the virus genome are carried out by host enzymes, or whether the virus P protein plays a role in some. The virus DNA is not replicated in the nucleus, but more copies are brought into the nucleus later in the replication cycle.

Transcription

The cccDNA is the template for transcription. You will recall that the HBV genome has four promoters, sited upstream of the pre-S1, pre-S2, X and pre-C regions. Only the pre-S2 promoter has a TATA box. At least two of the promoters are highly specific to liver cells and it has been shown that some of the transcription factors involved are liver cell proteins. This is at least part of the explanation why HBV is specific to liver cells. Rates of transcription from all four promoters are controlled by binding of cell transcription factors to two enhancer sequences present in the HBV genome. It is thought that the virus protein X is also a transcription factor, though it appears not to bind to DNA and so is probably not a typical transcription factor.

Transcription is carried out by the cell RNA polymerase II and results in the synthesis of four size classes of RNA. All the transcripts are capped at the 5' end and polyadenylated at the 3' end. They all use the same polyadenylation signal (TATAAA), so they have a common 3' end. Note that the 3.5 kb mRNAs are longer than the genome, which is 3.2 kb. During their synthesis part of the genome is transcribed twice, hence these RNAs have direct terminal repeats. In order to synthesize the 3.5 kb mRNAs the RNA polymerase must ignore the polyadenylation signal on the first pass. Studies with duck HBV indicate that this is controlled by a DNA sequence (called PET = positive effector of transcription), which somehow allows the RNA polymerase to continue through the polyadenylation signal on the first pass, but not on the second pass. On the first pass PET suppresses a sequence called NET (negative effector of transcription), which is needed for termination of transcription. If NET is deleted the RNA polymerase makes additional transits around the DNA, producing very long transcripts.

Translation

Of the RNAs that are approximately 3.5 kb long there are two subsets that differ slightly in size. The shorter subset, which does not include the start codon for the pre-C sequence, acts as mRNA for the C and P proteins. The C ORF is upstream of the P ORF, and most ribosomes that bind to these mRNAs commence translation at the C start codon.

There are at least three further start codons before the P start codon. When translation of P commences it appears that the ribosome bypasses the upstream start codons by 'leaky scanning'. Thus HBV produces much more capsid protein than polymerase, just as retroviruses control the relative amounts of Gag and Pol proteins synthesized. The shorter subset of 3.5 kb RNAs also acts as pregenomes.

HBeAg is translated from the longer subset of the 3.5 kb mRNAs, which include the start codon for the pre-C sequence. After translation HBeAg is secreted from the cell. There are also subsets of the 2.1 kb mRNAs. The M protein is translated from the longest subset and the S protein from several shorter subsets.

Post-translational modifications

Some of the envelope protein molecules become glycosylated and L is myristylated at the N terminus. It has been found that myristylation of L is not required for efficient virion assembly but virions produced in its absence are not infective.

Initially all the L protein N termini are on the cytoplasm side of the endoplasmic reticulum membranes, but after translation about 50 per cent of the N termini are moved through the membranes into the lumen of the endoplasmic reticulum. It is thought that the cell heat shock protein Hsc70 plays a role in this process.

Nucleocapsid assembly

The C protein forms dimers, which are assembled into capsids. It is known that the C protein can self-assemble, because capsids can be formed in *Escherichia coli* recombinants expressing the HBV C gene.

A molecule of P protein, along with several cell proteins, binds to a molecule of 3.5 kb RNA, which will function as pregenome RNA. P binds at a site within a sequence known as ε (Greek letter *epsilon*) that has a high degree of secondary structure. The ε sequence is within the terminal repeat and so is present at both ends of the

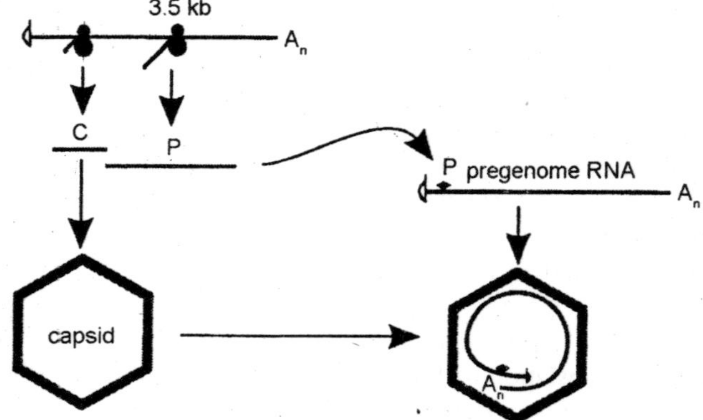

Fig. 14.13. Early stages of HBV assembly.

pregenome RNA. For some reason though, P binds only at the 5' end. The ε structure with its bound P protein acts as the packaging signal for incorporation of the pregenome RNA into a capsid.

Genome synthesis

The reverse transcriptase domain of P carries out DNA synthesis and the terminal protein domain of P acts as the primer for the initiation of minus-strand DNA synthesis. A covalent bond is formed between the -OH group of a tyrosine residue near the N terminus of P and the first nucleotide. There are conflicting reports as to whether the terminal protein becomes cleaved from the reverse transcriptase after the initiation of DNA synthesis.

The pregenome RNA acts as template for DNA synthesis. Initially a 4-nucleotide (–) DNA is synthesized then it is transferred to a complementary sequence in DR1 near the 3' end of the pregenome. DNA synthesis continues to the 5' end of the pregenome template. The RNase H activity of the P protein degrades the pregenome RNA from the RNA-DNA duplex. All the RNA is removed except for a remnant of 15–18 bases including the cap. The -OH group at the 3' end of the RNA remnant acts as the primer for the synthesis of (+) DNA.

During (+) DNA synthesis a nucleocapsid can either migrate to the nucleus to increase the pool of cccDNA or it can undergo a maturation event that enables it to bud through a membrane containing virus envelope proteins. DNA synthesis ceases on budding, as the nucleocapsid is cut off from the pool of nucleotides in the cytoplasm. This explains why the (+) DNA in the virion is incomplete.

Exit

The membranes through which budding occurs are part of a compartment between the endoplasmic reticulum and the Golgi complex. As budding proceeds copies of the L protein with their N termini on the cytoplasmic surfaces of the membranes bind to capsids. Virions are transported to the cell surface, where they are released from the cell. It is thought that transport occurs within vesicles, which fuse with the plasma membrane, releasing the virions by exocytosis. Non-infectious spheres and filaments are also released from the cell.

The host hepatocyte is not killed by HBV infection; the cell may survive for many months, releasing large quantities of virions and non-infectious particles. No cytopathic effect is observed in productively infected cell cultures (except for those infected with some HBV mutants). When the liver becomes damaged in an HBV-infected individual it is not as a result of cell damage caused by the virus, but it is due to the killing of HBV-infected cells by the body's immune system.

Overview of HBV replication cycle

Some aspects of this model have yet to be proven unequivocally and much interesting research remains to be done to unearth more details of the replication of this virus.

Prevention and Treatment of HBV Infection

The original HBV vaccines consisted of non-infectious spheres and filaments extracted from blood donations from HBV carriers. Now much vaccine is produced using recombinant yeast cells containing the gene for the S protein. Vaccination programmes have been successful in reducing the percentage of HBV carriers in some parts of the world such as Taiwan and Alaska. The requirements to mass produce HBV vaccine cheaply and to vaccinate many more of those at risk are major challenges facing the world.

Alpha-interferon has been used for some years to treat HBV-infected persons. This treatment does not eliminate the infection, but it results in a significant reduction in viraemia in about 20–30 per cent of cases. There is a price to pay for interferon treatment in the form of side-effects such as influenza-like symptoms and weight loss, which may necessitate reducing the dosage or even discontinuing the treatment.

The drug lamivudine, a nucleoside analogue, is also used to treat HBV infection. In many ways lamivudine is an improvement over α-interferon as it suppresses virus replication with a low incidence of side-effects, is administered by mouth rather than by injection and is cheaper. Longt-erm treatment, however, results in the appearance of lamivudine-resistant HBV mutants, though they appear to be susceptible to other nucleoside analogues such as adefovir.

Other reverse-transcribing DNA viruses

There are some plant viruses with dsDNA genomes that replicate by reverse transcription. These viruses are classified in the family *Caulimoviridae*, which includes viruses with isometric virions, such as cauliflower mosaic virus, and viruses with rod-shaped virions, such as rice tungro bacilliform virus.

HERPESVIRUSES

The herpesviruses derive their name from the Greek word *herpein*, meaning to creep. More than 100 herpesviruses have been isolated from a range of hosts that includes mammals, birds, fish, reptiles, amphibians and molluscs. Eight of these viruses are human viruses.

A notable characteristic of herpesviruses is that, once they have infected a host, they often remain as persistent infections for the lifetime of the host. These infections are often latent infections, which can be reactivated from time to time, especially if the host becomes immunocompromised. Both primary and reactivated herpesvirus infections can either be asymptomatic or can result in disease of varying severity. The outcome depends on the interplay between the particular virus and its host, and especially on the immune status of the host.

Human Herpesviruses

There are eight herpesviruses known in man, and most adults in the world are persistently infected with most of them.

Herpes simplex viruses 1 and 2

Herpes simplex viruses 1 and 2 (HSV-1 and HSV-2) initially infect epithelial cells of the oral or genital mucosa, the skin or the cornea. The virus may enter neurones and may be transported to their nuclei, where they may establish latent infections.

HSV-1 commonly infects via the lips or the nose between the ages of 6 and 18 months. A latent infection may be reactivated if, for example, the host becomes stressed or immunosuppressed. Reactivation results in the production of virions, which in about 20–40 per cent of cases are transported within the neurone to the initial site of infection, where they cause productive infection in epithelial cells, resulting in a cold sore. Occasionally there may be serious complications such as encephalitis, especially in immunocompromised hosts.

HSV-2 is the usual causative agent of genital herpes, which is a sexually transmitted disease. In newborn babies infection can result in serious disease, with a mortality rate of about 54 per cent.

Although the face and the genitals are the normal sites of infection for HSV-1 and HSV-2, respectively there are increasing numbers of cases where HSV-1 infects the genitals and HSV-2 infects the face.

Varicella-zoster virus

Infection with varicella-zoster virus usually occurs in childhood and causes varicella (chickenpox), when the virus spreads through the blood to the skin, causing a rash. It may also spread to nerve cells, where it may establish a latent infection. The nerves most often affected are those in the face or the trunk, and these are the areas most commonly affected in zoster (shingles) when a latent infection is reactivated.

Epstein-barr virus

Epstein-Barr virus (EBV) is transmitted in saliva. Epithelial cells are infected first then the infection spreads to B cells, which are the main host cell type for this virus. More than 90 per cent of people become infected with EBV, usually during the first years of life, when infection results in few or no symptoms. In developed countries some individuals do not become infected until adolescence or adulthood. A proportion of these individuals develop infectious mononucleosis (glandular fever), commonly called 'the *kissing disease*' by doctors. EBV is associated with a number of tumours in man.

Human cytomegalovirus

In the vast majority of infections with human cytomegalovirus symptoms are either absent or they are mild. In a pregnant woman, however, the virus can infect the placenta and then the foetus, for whom the consequences may be serious. In the US about one per cent of babies are born infected with the virus (about 40 000 per year). In about seven per cent of these there is evidence of virus-induced damage at birth, including small brain size and enlargement of the liver and spleen. In other individuals damage develops at a later stage; the damage may be manifest in a number of ways, such as hearing loss and mental retardation. Human cytomegalovirus can also cause severe disease (e.g. pneumonitis, hepatitis) in immunocompromised patients such as those with AIDS, those who have received treatment for cancer and those who are immunosuppressed because they have received an organ transplant.

Human herpesvirus 6

There are two types of human herpesvirus 6, known as HHV-6A and HHV-6B. Infection of a child with the latter can cause a fever and the sudden appearance of a rash known as exanthem subitum.

Human herpesvirus 7

Human herpesvirus 7 was first isolated from a culture of CD4 T cells that developed a cytopathic effect; the cells were from a healthy person. The virus has been associated with some cases of exanthem subitum.

Kaposi's sarcoma-associated herpesvirus

Kaposi's sarcoma-associated herpesvirus was discovered in 1994 and is named after the tumour with which the virus is associated.

Herpesvirus Virion

Herpesviruses have relatively complex virions composed of a large number of protein species organized into three distinct structures: capsid, tegument and envelope. The virus genome is a linear dsDNA molecule, which varies in size within the herpesvirus family from 125 to 240 kbp. The DNA is housed in the capsid, which is icosahedral, and the capsid is surrounded by the tegument. The HSV-1 tegument contains at least 15 protein species and some virus mRNA molecules. The envelope contains a large number of spikes (600–750 in HSV-1) composed of ten or more glycoprotein species. There are several different sizes of spike.

A number of schemes have evolved for the nomenclature of herpesvirus proteins, with the result that an individual protein may be referred to in the literature by two or more different names. Most of the structural proteins are commonly named VP (virus protein). In HSV-1 the most abundant proteins in the capsid and the tegument are VP5 and VP16, respectively. In the envelope there are at least 12 species of glycoprotein, each of which is prefixed 'g', for example gB, gC and gD.

The capsid is constructed from 162 capsomeres, 12 of which are pentons and the remainder of which are hexons. In HSV-1 the capsomeres are constructed from VP5: a penton is made from five molecules of VP5 and a hexon from six molecules. Other proteins make up structures called triplexes, which connect the capsomeres.

HSV-1 Genome Organization

HSV-1 is one of the most-studied herpesviruses; we shall look at its genome organization and then at its replication. The genome consists of two unique sequences each flanked by repeat sequences. The unique sequences are not of equal length: the longer is designated U_L and the shorter is designated U_S. The HSV-1 genome encodes at least 74 proteins plus some RNAs that are not translated. Both strands of the DNA are used for coding. The inverted repeats contain some genes, so the genome contains two copies of these genes, one in each strand.

HSV-1 Replication

Although HSV-1 infects only humans in nature, a variety of animal species and cell cultures can be infected in the laboratory. The replication cycle of the virus has been studied in cell cultures from a number of species, including humans, monkeys, mice and dogs.

Attachment and entry

The sequence of events at the cell surface usually involves the HSV-1 virion binding initially to heparan sulphate, and then to the main receptor. The latter can be one of several types of cell surface molecule including some nectins, which are cell adhesion molecules. The virion envelope then fuses with the plasma membrane. Infection may also occur by endocytosis, followed by fusion between the virion envelope and the endosome membrane. At least five of the glycoprotein species in the envelope are involved in these processes.

The nucleocapsid and the tegument proteins are released into the cytoplasm and the nucleocapsid must then be transported to the nucleus, where virus replication takes place. When the host cell is a neurone this journey is a long one; it has been estimated that it would take at least 200 years for the nucleocapsid to reach the nucleus by passive diffusion. In fact, the nucleocapsid is rapidly transported along microtubules to the vicinity of a nuclear pore. The virus DNA is released into the nucleus, where the linear molecule is converted into a covalently closed circular molecule, which becomes associated with cell histones.

Studies with antibodies specific for tegument proteins have provided evidence that these proteins are transported to several sites in the cell. At these sites tegument proteins play a variety of roles, including the down-regulation of host DNA, RNA and protein synthesis. One tegument protein known as virion host shutoff (vhs) protein degrades cell mRNA. Other tegument proteins are involved in the activation of virus genes, in particular the major tegument protein, VP16, which is transported to the nucleus, where it becomes associated with the virus DNA.

Transcription and translation

Herpesvirus genes are expressed in three phases: immediate early (IE), early (E) and late (L). Some authors refer to these phases by the Greek letters α, β and γ. There are introns in a few of the HSV-1 genes, mainly in the IE genes. The IE genes are activated by VP16. It was noted above that VP16 from infecting virions associates with the virus DNA. It does this by binding to a complex of cell proteins including Oct-1, which binds to the sequence TAATGARAT present in the promoter of each of the IE genes. VP16 then acts as a transcription factor to recruit the host RNA polymerase II and associated initiation components to each IE gene.

There are five IE proteins and all are transcription factors with roles in switching on E and L genes and in down-regulating the expression of some of these genes. At least some of the IE proteins have more than one role. Some of the E proteins have roles in virus DNA replication (next section), which takes place in discrete regions of the nucleus known as replication compartments. Virus DNA and proteins accumulate in these compartments along with cell RNA polymerase II, which transcribes the L genes. Most of the L proteins are the virus structural proteins.

Genome replication

The virus DNA is replicated by E proteins, seven of which are essential for the process. Copies of an origin-binding protein bind at one of three *ori* sites in the virus DNA. The protein has a helicase activity, causing the double helix to unwind at that site, and it has an affinity for the resulting single strands of DNA. The double helix is prevented from re-forming by the binding of copies of a ssDNA-binding protein.

The *ori* site is then bound by a complex of three proteins, which act as a helicase, further unwinding the double helix to form a replication fork. On one of the strands a complex of the same three proteins (now acting as a primase) synthesizes a short sequence of RNA complementary to the DNA. This RNA acts as a primer, and the DNA polymerase complexed with a polymerase processivity factor starts to synthesize the leading strand of DNA. On the other genome strand primases synthesize short RNAs, which are the primers for the synthesis of the Okazaki fragments of the lagging strand of DNA.

In addition to the activities of the seven proteins just described, other virus proteins, such as a thymidine kinase, may also be involved in DNA replication.

It is thought that the circular DNA is first amplified by θ replication, and that later the replication mode switches to σ, also known as rolling circle. The latter is the predominant mode of herpesvirus DNA replication, and the products are long DNA molecules known as concatemers, each of which consists of multiple copies of the virus genome.

Assembly and exit of virions from the cell

The virus envelope glycoproteins are synthesized in the rough endoplasmic reticulum and are transported to the Golgi complex. The other structural proteins, such as VP5, accumulate in the nuclear replication compartments, where procapsids are constructed. A procapsid is more rounded than a mature capsid and its structural integrity is maintained by the incorporation of scaffolding proteins. Before or during DNA packaging the scaffolding proteins are removed by a virus-encoded protease. Each procapsid acquires a genome-length of DNA, which is cut from a concatemer. Each cleavage occurs at a packaging signal at the junction of two copies of the genome. The DNA enters the procapsid via a portal at one of the vertices of the icosahedron. There is evidence from electron microscopy of capsids treated with antibody specific for pUL6 (the protein encoded by the *UL6* gene) that this protein forms the portal. It is likely that the DNA passes through the portal both when entering the procapsid during packaging and on leaving the capsid during the infection process, the portal playing a similar role to the head–tail connector in a tailed phage particle.

Once the nucleocapsid has been constructed it must acquire its tegument and envelope and then the resulting virion must be released from the cell; these are complex processes and many of the details are still unclear. It is thought that the sequence of events is as follows:

1. budding through the inner membrane of the nuclear envelope, giving the nucleocapsid a temporary envelope;
2. fusion of the temporary envelope with the outer membrane of the nuclear envelope, releasing the nucleocapsid into the cytoplasm;
3. acquisition of VP16 and other components of the tegument;
4. acquisition of the virion envelope by budding into a vesicle derived from the Golgi complex;
5. fusion of the vesicle membrane with the plasma membrane, releasing the virion from the cell.

Although the virion envelopes are derived from membranes within the cell, virus glycoproteins are also expressed at the cell surface.

These glycoproteins can cause fusion between infected cells and non-infected cells, resulting in the formation of giant cells known as syncytia, which can be observed both in infected cell cultures and in HSV lesions. Infected cell cultures are normally destroyed 18–24 hours after inoculation.

Latent herpesvirus infection

When infection of a cell with a herpesvirus results in latency rather than a productive infection, multiple genome is switched off during latency, but a few regions are transcribed and a few RNAs are synthesized; some viruses also synthesize a few proteins.

No virus proteins are required to maintain latency in cells that do not divide, so none are produced in neurones latently infected with HSV-1. Virus RNAs, however, are synthesized; these are known as latency- associated transcripts (LATs). Primary transcripts are synthesized from the *LAT* gene, which is located in the terminal repeats of the genome. The LATs undergo splicing, and at least one of them plays a role in inhibiting apoptosis, thereby ensuring the survival of the neurone with its latent HSV-1 infection. EBV, in contrast, becomes latent in memory B cells, which divide from time to time; the virus therefore synthesizes proteins needed to maintain the copy number of its genome when the host cell divides. The likelihood of a latent herpesvirus infection becoming reactivated is increased if the host becomes immunocompromised; the greater the degree to which the host is immunocompromised, the greater the likelihood of reactivation.

Other dsDNA viruses

Baculoviruses infect insects and other invertebrates. Some of them are used as insecticides and some are used as gene vectors in protein expression systems. Papillomaviruses are the causative agents of warts (papillomas) and some of them cause human cancers. Most dsDNA viruses, like the herpesviruses, encode genes in both strands of the DNA, so the strands cannot be designated as plus and minus. There are some dsDNA viruses, however, such as the papillomaviruses and phage T7, that encode all of their genes in one strand of the DNA.

15

Phylogeny of Viruses

In this chapter we shall speculate on how viruses may have originated and consider the mechanisms by which viruses evolved in the past and continue to evolve today. The origins and evolution of many cellular organisms can be inferred from fossils, but there is very little fossil record of viruses! The discussion of the possible origins of viruses will be highly speculative, though when we come to mechanisms of virus evolution we will be on firmer ground as sequencing of virus genomes enables the construction of phylogenetic trees and the monitoring of virus evolution as it occurs.

Origins of Viruses

Viruses by definition are parasites of cells, so there could be no viruses until cells had evolved. There is evidence for the presence of living cells about 3.9 billion years ago, though the first cells probably came into existence several hundred million years earlier. It is believed that, prior to the development of these primitive prokaryotes, there was a phase of evolution involving organic molecules. These molecules probably included proteins and RNA molecules, and some of the latter may have evolved capacities for self-replication.

The archaea and the bacteria that inhabit the earth today are the prokaryotic descendants of the early cells. It is likely that viruses developed at an early stage in these primitive prokaryotes, but the extent to which the viruses of the modern prokaryotes resemble these early viruses is unknown. Eukaryotic cells appeared much later in evolution, so on the evolutionary timescale the viruses that infect eukaryotes are much more recent than those of the prokaryotes. The viruses of modern mammals and birds are presumably descended from

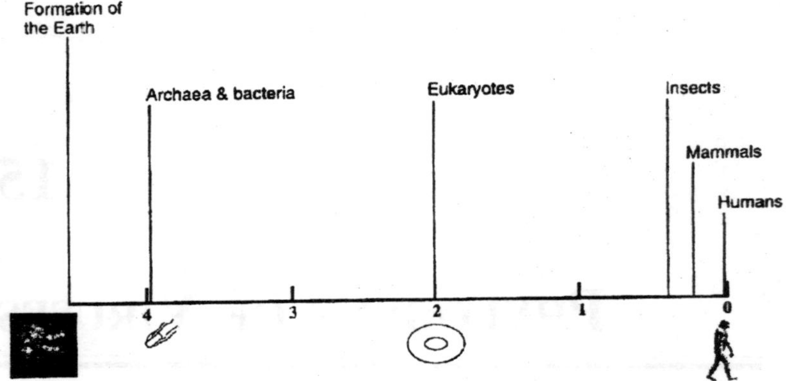

Fig. 15.1. Time of the Earth's history.

those that infected their dinosaur ancestors. The answer to the question 'Where did viruses come from?' is 'Basically, we do not know!'. We can, however, speculate about possible ancestors of viruses; these include molecular precursors of cellular organisms when the earth was young, components of cells and intracellular micro-organisms. We shall consider each of these in turn.

Molecular Precursors of Cellular Organisms

It is thought that on the early Earth, before cellular organisms had come into existence, RNA molecules evolved and developed enzyme activities (ribozymes) and the ability to replicate themselves. Once cells had evolved perhaps some were parasitized by some of these RNA molecules, which somehow acquired capsid protein genes. If this happened then these were the first viruses.

Components of Cells

Perhaps some cellular components evolved abilities to replicate themselves, independent of host cell control, and thus became parasites of those cells. Potential candidates for precursors of viruses include mRNA molecules and DNA molecules such as plasmids and transposons. The genomes of many plus-strand RNA viruses of eukaryotic hosts have characteristics of eukaryotic mRNA. For example, the 5' end of the genome is capped in tobamoviruses, the 3' end is polyadenylated in picornaviruses, while the genomes of coronaviruses are both capped and polyadenylated. The plus strands in the dsRNA genomes of reoviruses are also capped and polyadenylated. Could some RNA viruses be descendants of cell mRNAs?

Covalently closed circular DNA molecules known as plasmids are found in prokaryotic and eukaryotic cells. Could some DNA viruses be descendants of such molecules? Some bacterial plasmids carry genes that specify the production of a protein tube (a sex pilus) that can attach to other bacterial cells and allow the transfer of a copy of the plasmid from one cell to the other. Could some phages, such as the filamentous phages or some of the tailed phages, be descended from ancient plasmids and their sex pili? On the other hand it has been suggested that the sex pili of modern bacteria may have originated from filamentous phages!

Transposons are sequences of mobile DNA in the genomes of prokaryotes and eukaryotes. They are described as mobile because they can move from one part of the genome to another by 'cut and paste' and 'copy and paste' mechanisms. Perhaps some DNA viruses are descendants of transposons.

For any of these DNA or RNA molecules to evolve into a virus the putative virus genome would somehow have to acquire a range of genes, including those for capsid protein(s) and, in many cases, for a polymerase to replicate the virus genome.

Another class of cellular components that are strong contenders as precursors of viruses are retrotransposons, which are retrovirus-like sequences in eukaryotic cell genomes. A retrotransposon consists of two long terminal repeats (LTRs) flanking a *gag* gene, a *pol* gene (encodes a reverse transcriptase and an integrase) and sometimes an *env* gene. Retrotransposons thus resemble the proviruses of retroviruses; they can replicate by transcribing mRNAs, which are translated into Gag and Gag–Pol proteins. Reverse transcription produces a new copy of the retrotransposon, which can be inserted into another site in the cell genome. Retrotransposons are found in the genomes of humans, other vertebrates, invertebrates, plants and fungi.

It is likely that retroviruses originated from retrotransposons, though this is a chicken-and-egg situation and it is also likely that some retrotransposons are descended from retroviral proviruses that became inserted into the germ line of their hosts. Retrotransposons are described as autonomous elements because they have a *pol* gene that enables them to replicate.

The human genome also contains many non- autonomous elements flanked by LTRs. Some of these elements consist only of *gag* sequences, which could be remnants of either retrotransposons or retroviral proviruses.

Results from the human genome sequencing project indicate that about eight per cent of our genome is composed of LTR-flanked sequences, which are often referred to as endogenous retroviruses. David Baltimore (one of the discoverers of reverse transcriptase) has said 'in places, the genome looks like a sea of reverse-transcribed DNA with a small admixture of genes'. Other reverse transcribing viruses, such as hepadnaviruses, might have similar origins to retroviruses.

Intracellular Micro-organisms

There is strong evidence that the mitochondria and chloroplasts of eukaryotic cells are derived from prokaryotic cells that adopted new modes of life within the cells of host organisms. It is assumed that the ancestors of these organelles adopted parasitic or mutually beneficial modes of life in host cells, and that over time they became increasingly dependent on their hosts, losing the ability to perform various functions, and losing the genes that encode those functions. Perhaps a similar evolutionary process continued further, leading to greater degeneracy and loss of functions such as protein synthesis, until the intracellular intruder was no longer a cell or an organelle, but had become a virus.

A virus that may have arisen in this way is the mimivirus. The 1.2 Mb genome of this virus encodes a wide variety of proteins, including enzymes for polysaccharide synthesis and proteins involved in translation; it also encodes six tRNAs.

How Did Viruses Originate?

The answer to this question is that we do not know! The great variety of virion structures and of virus genome types and replication strategies indicates that viruses had multiple origins. Small simple viruses, such as parvoviruses and picornaviruses, perhaps evolved from molecular precursors, while some large complex viruses, such as the mimivirus, perhaps evolved from cellular precursors.

We noted that the viruses found in certain categories of organism have a predominance of certain types of genome: dsDNA genomes in prokaryote viruses, ssRNA genomes in plant viruses and dsRNA genomes in fungal viruses. We can now note that certain structural types of virus are restricted to particular categories of host: naked, rod- shaped ssRNA viruses are restricted to plant hosts, while viruses with a head-plus-tail structure are rarely found outside the prokaryotes. The reasons for these distributions presumably concern diverse origins of the viruses in these very different hosts. There is one feature, however, that is universal throughout the virus world: icosahedral symmetry. There are icosahedral viruses with dsDNA, ssDNA, dsRNA and ssRNA genomes, and icosahedral viruses infect bacterial, fungal, plant and animal hosts. These viruses have evolved from a variety of origins, but evolution has chosen the icosahedron for capsid design because it is a particularly stable structure. One final thought on virus origins: we must entertain the possibility that today new viruses are continuing to evolve from molecular and cellar precursors.

Evolution of Viruses

Charles Darwin's concepts of over-reproduction and survival of the fittest that have been found to apply to cellular organisms also apply to viruses. Like a cellular organism, a virus has genes that strive to perpetuate themselves. The expression of the combination of genes that constitutes the genome permits the virus to replicate itself. Changes to individual genes and new gene combinations continually throw up new genotypes, most of which are less successful than the parental genotypes and do not survive. Occasionally, however, a new genotype is more successful than the parental genotype and might even supplant it. Sometimes a new virus genotype enables the infection of a new host species.

Viruses from the Past

The ancestors of modern viruses ('virus fossils') are not readily available, but the reverse transcription–polymerase chain reaction (RT-PCR) has been used to recover and amplify genomes of at least two RNA viruses from the past: influenza A virus from the 1918–19 pandemic and tomato mosaic virus. The latter virus is very stable and it can be detected in the atmosphere. Its presence in Arctic ice was predicted, and when cores of ice between 500 and 140 000 years old were examined the RNAs of 15 strains of tomato mosaic virus were detected.

Monitoring Virus Evolution

Comparing sequences of virus genomes allows the degrees of evolutionary relationship between viruses to be estimated, and even allows changes in rapidly evolving viruses (e.g. HIV-1) to be followed. Sequence data can be used to produce phylogenetic trees, which indicate how modern virus strains may have arisen from common ancestors.

Mechanisms of Virus Evolution

In many respects, the underlying processes that drive virus evolution are the same as those that drive the evolution of cellular

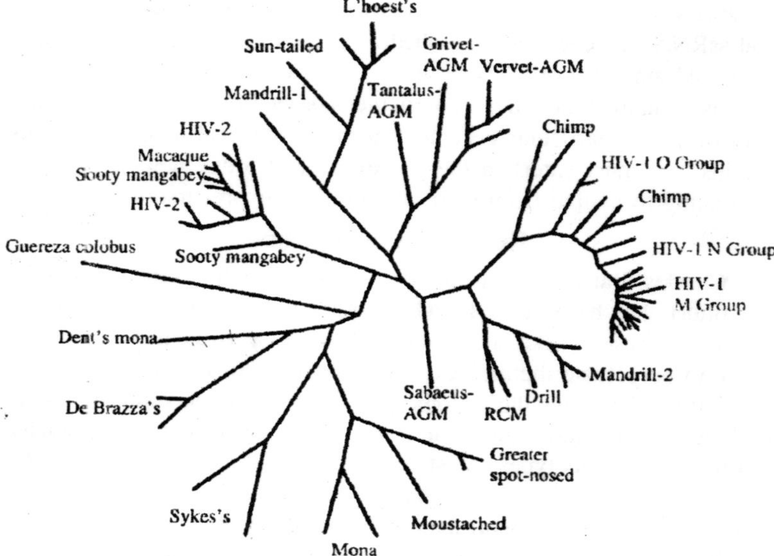

Fig. 15.2 Phylogenetic tree showing relationships between primate lentiviruses based on pol sequences.

organisms. These processes involve the generation of genome variants, the vast majority of which are deleterious and do not survive, but a few provide an advantage in a particular niche. For a virus the niche might be a new host species or the presence of an antiviral drug, and a variant may proliferate in that niche as a new virus strain. Virus genome variants arise as a result of mutations, new gene arrangements and acquisition of cell genes; we shall consider each process in turn.

Mutations

When nucleic acids are copied by polymerases some errors are made. If an error is in a protein-coding sequence and if the error results in a change in the amino acid encoded, then the error results in a mutation. Natural selection operates and the mutations that survive are those that best fit the virus for its continued survival.

There are many selection pressures on viruses. These pressures include the immune response of the host; for example a new antigenic type of an animal virus that the host's immune system has not previously encountered is at an advantage compared with antigenic types against which the host has acquired immunity. There is, therefore, heavy selection pressure on virus proteins (e.g. HIV-1 gp120) that are targets of the host's immune response, and these proteins tend to be the least conserved. Those regions of the proteins that are targets for neutralizing antibodies are especially variable. There are, however, constraints on the evolutionary process. Virus attachment proteins must retain the configurations that enable them to bind to cell receptors, and enzymes, such as reverse transcriptases, must retain their catalytic abilities.

For some viruses there are additional constraints. A virus such as potato yellow dwarf virus, which needs to replicate in both a plant host and an insect host, can tolerate a mutation that enhances its replication in the insect host only if the mutation does not compromise its ability to replicate in the plant host, and *vice versa*.

When virus DNA genomes are replicated, the error rates are much lower than when RNA genomes are replicated. This is because DNA- dependent DNA polymerases have a proofreading mechanism that corrects most of the errors, whereas RNA polymerases and reverse transcriptases have no such mechanism. As a result, DNA viruses, such as papillomaviruses, evolve much more slowly than RNA viruses, such as picornaviruses.

The high error rates during RNA replication and reverse transcription mean that for an RNA virus there is no fixed sequence of bases for the genome. Instead, the virus genome exists as a large number of variants; the term 'quasispecies' has been coined to describe the group of variants that collectively constitute the genome of an RNA virus. Many variants have only a fleeting existence, while those best adapted to a particular niche dominate in that niche. The high mutation rate produces some variants that may enable a virus to adapt to new niches, for example an expansion of its host range, should the opportunities arise.

The first step in the replication of HIV genomes involves reverse transcription, with an estimated error rate around one error per 10^4 bases added. As the HIV genome contains about 10^4 bases, this means that there is an average of one error each time the genome is reverse transcribed. There is evidence that the human immunodeficiency viruses evolved from viruses that jumped from chimpanzees and sooty mangabeys into humans. Since jumping species, the viruses have evolved rapidly, especially the M (main) group of HIV-1, which has radiated into many subtypes and sub-subtypes.

Rapid evolution of HIV can be detected within the body of an infected individual, in which there may be 10^{10}–10^{12} new virions produced each day. There is therefore constant production of a huge amount of genetic variability. We mentioned previously that R5 strains of HIV-1 predominate in a newly infected host, but in many individuals as the infection proceeds there is evolution towards X4 and R5X4 strains. The constant production of new variants also means that the virus can evolve rapidly in response to a selection pressure, such as the presence of an anti-retroviral drug.

High diversity is also found in the genomes of other RNA viruses, such as hepatitis C virus, where isolates from a patient can have up to five per cent difference in their nucleotide sequences.

The capacity for rapid evolution of RNA viruses creates problems for the development and maintenance of some vaccination programmes. Vaccination with attenuated poliovirus vaccine initiates infection in the gut with one or more of the vaccine strains, and the infection normally lasts for 1–2 months. Some immunodeficient vaccinees, however, remain infected for much longer periods of time, and in these individuals the vaccine viruses almost always evolve. There are differences of only a few bases between the genomes of neurovirulent poliovirus strains and those of the vaccine strains, so reversion to neurovirulence is common in these chronically infected individuals, who may develop paralytic poliomyelitis and may transmit virulent poliovirus to others.

Another example of RNA virus variability having practical consequences for vaccination is foot and mouth disease virus. Seven distinct serotypes of this virus have evolved, and vaccination against one serotype does not provide protection against the others. In addition, there are many subtypes of each serotype and immunity to one subtype confers only partial immunity to the related subtypes.

Hepatitis B virus (HBV) is a DNA virus, but, because it replicates its genome by reverse transcription, error rates similar to those of RNA viruses are observed. More than half of the HBV genome is read in two reading frames and regulatory elements are all within protein coding regions. There are, therefore, severe constraints on the ability of HBV to mutate and remain viable but, in spite of these constraints, variation is much greater than in DNA viruses that replicate their genomes solely with DNA- dependent DNA polymerases.

Recombination

Recombination is a process that results in the production of a new genome, derived from two parental genomes. For a cellular organism recombination can take place between DNA molecules within a cell. For viruses recombination may occur when a cell is infected with two related viruses; the new virus that is produced is referred to as a recombinant virus. Both DNA and RNA viruses can undergo recombination.

The first evidence for recombination in a virus came from studies with the DNA phage T2, and since then recombination has been documented for other DNA viruses. The formation of recombinants between different baculoviruses (DNA viruses that infect invertebrates such as insects), and between herpes simplex virus 1 and herpes simplex virus 2, has been reported. The recombination mechanism operating for DNA viruses is cleavage of the parental DNA molecules followed by ligation. The genomes of RNA viruses may be able to undergo recombination by a similar mechanism. For ssRNA viruses such as picornaviruses, it is thought that recombinant genomes can also be produced by a template switching mechanism. This involves displacement of an RNA polymerase transcribing a molecule of RNA, followed by association of the enzyme with another RNA molecule and resumption of RNA synthesis using the new template.

Recombination appears to be a common event in retroviruses and pararetroviruses. It probably occurs during reverse transcription by a template switching mechanism similar to that described above, with the reverse transcriptase switching between two RNA templates. The RNA templates for a retrovirus are in the infecting virion, derived from two virus strains co-infecting the cell in which the virion was assembled. For a pararetrovirus the RNA templates are present in the cytoplasm of a cell co-infected with two strains. Many recombinant strains of HIV-1 have evolved; for example, a recombinant derived from HIV-1 subtypes A and E has become prevalent in Thailand.

Recombination occurs in nature, but recombinants with desired combinations of genes can be generated in the laboratory. Examples are the recombinant vaccinia virus that was produced for vaccination of foxes and recombinant baculoviruses that are produced for use in gene expression systems.

Reassortment

Reassortment is a category of recombination that may occur with segmented genome viruses that have all the segments packaged in one virion, for example reoviruses, bunyaviruses and influenza viruses. If a cell is co-infected by two strains of one of these viruses the progeny virions may contain mixtures of genome segments from the two parental strains. These virions are known as *reassortants*.

The virus genome is composed of eight segments of minus-strand RNA; two of the segments encode the envelope glycoproteins haemagglutinin (H) and neuraminidase (N). The figure shows how reassortment gives rise to a virus strain with a new H gene. During the 20th century there were three pandemics of influenza caused by new reassortants of influenza A virus. The most devastating pandemic occurred in 1918–19, when the number of people killed by the virus was greater than the number killed during the First World War. Using RT-PCR, influenza virus RNA sequences have been recovered from the tissues of some victims. In one project lung tissue that had been fixed in formalin and embedded in paraffin was the source, while another project used lung tissue from a body that had been buried in permafrost in Alaska. The influenza virus sequences that were found differ from any H1N1 viruses known today and the origins of the 1918 virus remain obscure.

Using reverse genetics it has been possible to clone the genes of the 1918 virus. This has enabled studies on the roles of individual genes in virus pathogenicity. It was shown that the H gene of the 1918 virus conferred pathogenicity to human influenza A virus strains that were non-pathogenic in mice. The origins of the viruses responsible for the two more recent influenza pandemics have been traced. In each case the virus was a reassortant between a human virus and an avian virus; the 1957 reassortant derived the H and N genes from an avian virus, while the 1968 reassortant derived the H gene from an avian virus.

Acquisition of cell genes

The discovery of similarities between some virus proteins and cell proteins has led to the suggestion that some viruses have 'captured' cell genes. The genomes of some retroviruses contain oncogenes that they probably acquired from their host cells, which contain similar genes (proto-oncogenes). An example is the *src* gene of Rous sarcoma virus; a very similar gene (c-*src*) is present in the cells of all vertebrates. A retrovirus might acquire an oncogene by recombination between the proviral DNA and host DNA, or by RNA synthesis during transcription continuing from the provirus through into a protooncogene adjacent in the cell chromosome.

Some virus proteins modulate the immune system of the host, often by mimicking cell proteins. For example, some of the human herpesviruses encode proteins similar to cytokines and major histocompatibility complex proteins. Several large DNA viruses encode proteins similar to interleukin-10 and some of these proteins have been shown to suppress cell- mediated immune responses. Are some of the virus genes that encode these proteins derived from cell genes and, if so, how did the viruses acquire the genes?

One possible mechanism is recombination, though in order for this to occur the genomes of the virus and the cell must come into close proximity. This may happen with DNA viruses that replicate in the nucleus (e.g. herpesviruses), but there are many DNA viruses (e.g. poxviruses) that do not enter the nucleus, so cellular and viral genomes are never in close proximity.

A second possible mechanism for viral acquisition of a cell gene might be through synthesis of a DNA copy of a cell mRNA, followed by insertion of the DNA into the virus genome. The DNA synthesis would require a reverse transcriptase, possibly supplied by the cell or by a retrovirus co-infecting the cell.

Evolution of New Viruses

From time to time a new virus appears, and mystery surrounds its origin. Often there are suspects for the origin of the virus and these suspects may be viruses in other host species. The genomes of the new virus and the suspects can be sequenced, and if a close match is found between the new virus and one of the suspects then the mystery has probably been solved. It is commonly found that a new virus is one that has infected a new host species in which some further evolution has then taken place.

Many virus infections across species barriers are dead ends for the viruses as no transmissions to further individuals occur, but sometimes a virus adapts to enable transmission from host to host. A virus may have to undergo one or more adaptations in order for it to

1. replicate in the new host;
2. evade immune responses of the new host;
3. transmit to other individuals of the new host.

In 1978 a new parvovirus appeared in dogs and rapidly spread around the world. The genome sequence of this canine parvovirus was found to be more than 99 per cent identical to that of a parvovirus in cats (feline panleukopenia virus). Both viruses can infect feline cells, but only the canine virus can infect canine cells. The feline virus is thought to be the precursor of the canine virus.

Similarly, the genome sequences of HIV-1 and HIV-2 are very similar to those of simian immunodeficiency viruses found in chimpanzees (SIV_{cpz}) and sooty mangabeys (SIV_{sm}), respectively. The human viruses are believed to have originated when these simian viruses jumped species, perhaps when humans came into contact with the blood of infected animals.

Co-evolution of Viruses and their Hosts

A virus–host association that has existed for a long period is likely to have evolved a relationship in which the host suffers little or no harm. Examples of such viruses are the dependoviruses, and some reoviruses, which acquired the 'o' in their name because they were found to be 'orphan' viruses. The members of these virus groups that infect *Homo sapiens* have probably been with us since we diverged as a separate species.

When a virus extends its host range into a new species it is often much more virulent in the new host than in the old. SIV_{cpz} and SIV_{sm} are both avirulent in their simian hosts, but their derivatives, HIV-1 and HIV-2, are lethal to humans. Of course the potential for evolution is usually much greater for a virus than it is for its host. As noted earlier, the reproductive capacity of HIV (10^{10}–10^{12} new virions per day) dwarfs that of the host.

A virulent virus may affect the evolution of its host by causing genes for susceptibility to infection with the virus to become extinct, whilst selecting in favour of genes for resistance. HIV-1 is currently selecting for the 32-nucleotide deletion in the *CCR5* gene that confers resistance to infection.

A good example of co-evolution of a virus–host relationship is myxoma virus in the European rabbit. In the natural host of the virus, a rabbit species in South America, infection results in the development of skin tumours, which eventually heal. In contrast, when myxoma virus infects the European rabbit it causes myxomatosis, characterized by acute conjunctivitis, lumps on the skin and loss of appetite. The disease is almost always fatal, and for this reason the virus was introduced into Australia as a biological control agent against the European rabbit. Within a year of its introduction there was evidence of virus evolution. Myxoma virus strains with attenuated virulence had evolved, apparently selected because they were transmitted more effectively. Subsequently, selection for rabbits with resistance to myxomatosis occurred and both the virus and its new host have continued to co-evolve.

The ultimate in virus–host co-evolution occurs if a virus genome becomes permanently integrated into the genome of its host. This is probably what happened to create at least some of the endogenous retrovirus/retrotransposon sequences in the genomes of eukaryotes.

The genomes of prokaryotes too contain sequences that originated as a result of integration of viral sequences through the process of lysogeny. Phage sequences (prophages) have been found in the majority of the bacterial genomes that have been sequenced. Some of these prophages are inducible, meaning that they can be activated, virus replication can ensue and the host can be lysed. Other prophages are defective and replication cannot be induced. The genes for some characteristics of bacteria are encoded by prophage sequences, for example, the cholera toxin gene of *Vibrio cholerae* and toxin genes in *E. coli* O157.

EMERGING VIRUSES

The term 'emerging virus' is used in a number of contexts: it may refer to a virus that has recently made its presence felt by infecting a new host species, by appearing in a new area of the world or by both. Sometimes a virus is described as a 're-emerging virus' if it has started to become more common after it was becoming rare. Foot and mouth disease virus re-emerges in the UK from time to time.

Human activities may increase the likelihood of virus emergence and re-emergence. Travel and trade provide opportunities for viruses to spread to new areas of the planet, and the introduction of animal species into new areas, e.g. horses into Australia, may provide new hosts for viruses in those areas.

Other activities that may result in virus emergence involve close contact with animals, including the hunting and killing of non-human primates for bushmeat. It has been shown that viruses such as simian immunodeficiency viruses (SIVs) can be present in the meat and there is a risk of acquiring an infection when the meat is handled. Simian-to-human transmission of SIVs is thought to have occurred several times, resulting in the major groups of HIV-1 and HIV-2, and there is concern that further viruses might emerge as a result of contact with bushmeat.

If a virus jumps into a new host species it may undergo some evolutionary changes in the new host, resulting in a new virus. This is how HIV-1 and HIV-2 were derived from their SIV precursors. Other new viruses emerge when recombination and reassortment result in new viable combinations of genes; new strains of influenza A virus come into this category. Some 'new' viruses that are reported are actually old viruses of which mankind has recently become aware, such as Kaposi's sarcoma-associated herpesvirus.

Viruses in New Host Species

Bunya viruses

In 1993 in a region of the south-western US known as Four Corners some residents became afflicted by an illness resembling influenza; many of them developed severe lung disease and died. The region is normally very dry, but there had been unusually high rainfall and snowfall that had resulted in a burst of plant growth followed by explosions in the populations of small mammals. One of those mammals was the deer mouse, a species that is attracted to human habitation.

On investigation, it was found that many of the deer mice were persistently infected with a virus that was excreted in their urine, droppings and saliva. Humans exposed to these materials were becoming infected and developing the respiratory illness. The virus was characterized as a new hantavirus and the disease was called hantavirus pulmonary syndrome.

Hantaviruses are members of the family *Bunyaviridae* and are named after the River Hantaan in Korea, where the first of these viruses was isolated during the Korean War from soldiers who had developed haemorrhagic fever with renal syndrome. Similar viruses are known elsewhere in Asia and in Europe.

There was much heated debate as to what name should be given to the virus that had appeared in Four Corners. Local people did not want it named after a place as they did not want potential tourists to be discouraged by publicity surrounding a virus disease. Eventually the Spanish name Sin Nombre virus (the virus without a name) was agreed.

Since 1993 hantavirus pulmonary syndrome, caused by Sin Nombre virus and other hantaviruses, has been reported in many parts of North America and South America, with mortality rates around 50 per cent.

Paramyxoviruses

Hendra virus

In 1994 at Hendra in south-east Australia there was an outbreak of pneumonia in horses, then a trainer and a stable hand who worked with the horses developed severe respiratory disease. The horses and the two humans were found to be infected with a virus that had characteristics of the family *Paramyxoviridae*. This family contains well known viruses such as measles and mumps viruses, but the virus that was isolated was previously unknown. Thirteen of the horses and the trainer died as a result of infection with the virus, which was named Hendra virus.

A couple of years later a man who lived about 800 km north of Hendra developed seizures and paralysis. He was found to be infected with Hendra virus and he died from meningoencephalitis. It turned out that he had helped with post mortems of two horses 13 months earlier. Tissue from the horses had been preserved and when it was examined it was found to contain Hendra virus.

It looked as though the infected humans had acquired the virus from horses, but the source of infection for horses was a mystery. An investigation was initiated, and evidence was obtained that the virus is present in populations of fruit-eating bats; Hendra virus-specific antibodies were found in all four local species of the bat genus *Pteropus*, and infectious virus was isolated from one bat. Bats appear to be the normal hosts of Hendra virus, and experimental infection of several bat species has provided no evidence that infection causes disease in these animals. It appears that occasional transmission to horses occurs, with very occasional transmission from horses to humans. In these species the outcome of infection is very different, with the virus demonstrating a high degree of virulence.

Nipah virus

In 1997 a new disease of pigs appeared in Malaysia. Affected animals developed respiratory disease and encephalitis. Soon afterwards workers on the affected pig farms began to develop encephalitis; over a two- year period there were several hundred human cases, with over 100 deaths. In 1999 a new virus was isolated from the brain of one of the patients who had died. The virus turned out to be a paramyxovirus with similar characteristics to Hendra virus and it was named Nipah virus. In an attempt to contain the outbreak over one million pigs were slaughtered.

A search was initiated to find the reservoir of infection. As with Hendra virus, bats were investigated and evidence of Nipah virus infection was found in fruit-eating bats of the genus *Pteropus*. Infectious virus was isolated from bats' urine and virus-specific antibodies were found in their blood. Thus the situation parallels that of Hendra virus, with a reservoir of infection in bats and transmission to humans via an intermediate mammalian host. More recently there have been outbreaks of Nipah virus encephalitis in humans in Bangladesh and India. In these outbreaks there was no evidence of transmission from pigs and it is believed that transmission may have occurred from bats to humans. There was also evidence of human-to-human transmission of the virus.

Viruses in New Areas

West nile virus

In 1999 some people in New York became ill with apparent viral encephalitis; several of them died. At about the same time several birds in the zoo became ill, and deaths of wild birds, especially crows, were reported. Diagnostic tests revealed that these human patients and the birds were infected with West Nile virus (WNV).

This virus has been known since 1937 when it was first isolated from the blood of a woman in the West Nile region of Uganda. In addition to Africa the virus is also known in Eastern Europe, West Asia and the Middle East, but until 1999 it had never been reported in North America. Most infections of humans and animals with WNV are asymptomatic, but about 20 per cent of human infections result in signs and symptoms such as fever and aching muscles. Less than one per cent of infections spread to the central nervous system, causing severe illness (such as encephalitis), which may result in death; these cases are mainly in those over 60 years old.

WNV is classified in the family *Flaviviridae*. The virus has a wide host range and is common in birds, humans and other vertebrates; it also replicates in mosquitoes, which act as vectors between the vertebrate hosts. Protective measures include use of insect repellents, wearing 'mosquito jackets' and treating mosquito breeding areas with insecticides. It is not known how WNV was introduced into North America, but once there it spread rapidly throughout the US; within five years it was also present in Canada, Central America and islands in the Caribbean. The virus has caused hundreds of human deaths in North America.

The spread of WNV to North America prompted an investigation to determine whether the virus is present in birds in the UK. WNV-specific antibodies were found in a number of species and, using RT-PCR, WNV RNA was detected in magpies and a blackbird, though all the birds appeared healthy.

Bluetongue virus

Bluetongue virus (BTV), a member of the genus *Orbivirus* in the family *Reoviridae*, causes disease in ruminant animals, including sheep, goats and cattle. Like WNV it has a wide host range in vertebrate animals and replicates in insects, which act as vectors.

All ruminant species are susceptible to infection with BTV, but infection results in severe disease and death only in certain hosts, mainly some breeds of sheep and some species of deer. In some affected animals the tongue may become swollen and blue. Infected animals that remain healthy can act as reservoirs of infection. The virus is transmitted between its ruminant hosts by certain species of biting midge in the genus *Culicoides*, so the virus is restricted to areas of the world where these midge species occur, and transmission is restricted to those times of the year when the midges are active.

Bluetongue has been known in Africa and the Middle East for a long time. Occasionally it has made brief incursions into Europe, but in 1998 six serotypes of BTV began to spread north. By 2004 there were cases 800 km further north than ever before and over a million sheep had been killed by the disease. It has been found that the midge species that is the main BTV vector has extended its range north in Europe, presumably because of the climate becoming warmer, and this is a major factor responsible for the spread of the virus.

Viruses in New Host Species and in New Areas

Filoviruses

In 1967 in the town of Marburg in Germany some laboratory workers became ill with a haemorrhagic fever (a terrible disease characterized by diarrhoea and vomiting, as well as by haemorrhaging and fever). These people had been in contact with blood, organs and cell cultures from African green monkeys caught in Uganda. Seven of those affected died, and there were five cases of the illness in hospital staff that had been in contact with patients' blood. Investigations revealed that the monkeys had been infected with a virus that had been transmitted to the laboratory staff and from them to the hospital staff. This virus was of a type never previously encountered, with elongated virions, some of which were straight and some were curved. It was named Marburg virus and is now classified in the family *Filoviridae*, named from the Latin *filum*, meaning a thread.

In 1976 there were outbreaks of a similar disease in Africa near the River Ebola in the Democratic Republic of Congo (then Zaire) and in Sudan. A virus similar to Marburg virus was isolated from patients and was named Ebola virus. Since then there have been a number of outbreaks of disease caused by Ebola and Marburg viruses across central Africa, from Cote d'Ivoire in the west to Kenya in the east and to Angola in the south.

The way in which the outbreaks start has long been a mystery. There is increasing evidence that some outbreaks start when a human becomes infected as a result of contact with the blood of an infected non-human primate. It was known from the original outbreak in Germany that African green monkeys can be infected with Marburg virus. It has since been found that gorillas, chimpanzees and duikers can be infected with Ebola virus, which may be responsible for significant mortality of these species. Ebola and Marburg viruses are present in the blood of infected hosts and transmission to humans can occur through contact with the flesh of infected animals after they have been hunted and killed. Human-to-human transmission readily occurs through contact with the blood of infected individuals.

There are several species of Ebola virus in Africa. A further species has appeared in the US and Italy in monkeys imported from the Philippines; there is serological evidence that animal handlers have been infected, but there have been no reports of this virus causing disease.

It is thought likely that additional animal species act as 'reservoirs' of Marburg and Ebola viruses, and that infection in these species is likely to cause few or no signs of disease. There have been many expeditions to search for reservoir hosts, but very few produced positive results; Ebola virus RNA has been detected in some small mammals including fruit bats, and anti-Ebola virus antibodies have been detected in some fruit bats. There remain unanswered questions concerning reservoir species and host ranges of these viruses.

Monkeypox virus

Monkeypox is a disease of monkeys that is similar to smallpox in man. The viruses that cause these diseases are members of the family *Poxviridae*. Since 1970 there have been cases in Africa of human disease caused by infection with monkeypox virus; some cases have been fatal. In 2003 some people in the US developed a mystery illness that involved a rash and a fever. Investigations revealed that they were infected with monkeypox virus, which had entered the country in

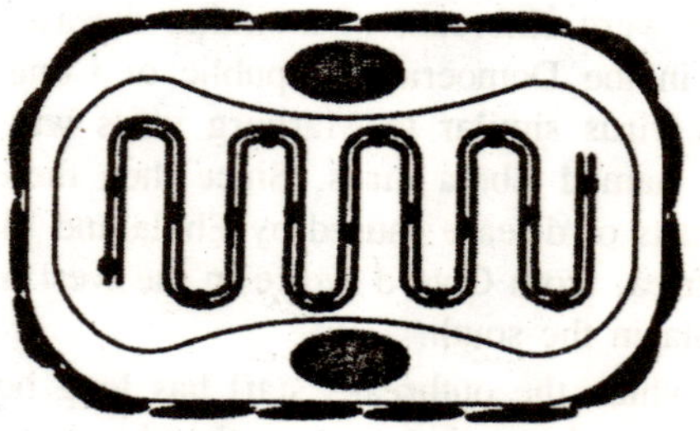

Fig. 15.3. Family poxviridae.

small mammals imported from Ghana. The virus had infected pet prairie dogs and had then been transmitted to humans. Sequencing the genomes of monkeypox virus isolates has revealed the existence of two strains that appear to differ in virulence. The virus that was imported into the US was the less virulent strain.

New Viruses

SARS coronavirus

In 2002 a new human respiratory disease emerged in southern China. The following year one of the doctors who had been treating patients travelled to Hong Kong, where he became ill and died. Subsequently, people who had stayed in the same hotel as the doctor travelled to Singapore, Vietnam, Canada and the US, taking the infectious agent with them. The epidemic of severe acute respiratory syndrome (SARS) was under way.

The signs and symptoms of SARS resemble those of influenza and include fever, aching muscles, sore throat, cough and shortness of breath. About 90 percent of patients recovered, but for the remainder the infection proved to be lethal. These were mainly individuals who had an underlying condition such as diabetes, heart disease or a weakened immune system. On the face of it SARS is a respiratory tract disease, but in many patients the infection spread to other parts of the body. Diarrhoea developed in some patients and the virus was shed in the faeces and urine for several weeks.

The causative agent was found to be a new coronavirus. No natural reservoir for the SARS coronavirus has been found, but coronaviruses with very similar genome sequences were isolated from animals sold in markets in the region of China where the first SARS cases appeared. Antibodies to these viruses were found in many workers in the markets, but none of the workers had any history of SARS. A similar virus has also been isolated from several bat species. It is likely that coronaviruses have repeatedly crossed into humans from other mammalian species, but in the majority of cases the virus has lacked the ability for efficient human-to-human transmission. It would seem that the SARS coronavirus evolved from a rare virus that had this ability. The SARS outbreak was brought under control by quarantine measures, but only after there had been over 8000 cases with nearly 800 deaths.

New strains of influenza virus A

The family *Orthomyxoviridae* includes the influenza A viruses. The virions have two species of surface glycoprotein: a haemagglutinin

(H) and a neuraminidase (N). There are 16 types of H antigen and nine types of N antigen, and there are many subtypes of each type. From time to time a virus emerges with a new combination of H and N genes formed by reassortment, and causes a pandemic.

The hosts of influenza A viruses are principally birds that frequent aquatic habitats. The birds (e.g. ducks, geese, gulls) acquire infections by ingestion or inhalation and the viruses infect their intestinal and/or respiratory tracts. Infection with most virus strains results in few or no signs of disease, but some strains are highly pathogenic and can kill their avian hosts. The viruses can be spread to new areas when the birds migrate. Some influenza A viruses infect mammalian species including pigs, horses and humans; the respiratory tract is the main site of virus replication. Normally humans are infected only with viruses that have H type 1, 2 or 3 and N type 1 or 2. Patients are commonly very ill and some die, either as a direct result of the virus infection, or indirectly from secondary pathogens, which are able to infect as a result of damage to the respiratory epithelium.

The previous two paragraphs describe the normal situation, but exceptions occur: some avian strains of influenza A virus can be highly pathogenic in birds, can be transmitted from wild birds to domestic poultry and can be transmitted to humans. This was the situation in Hong Kong in 1997, when an H5N1 virus caused an outbreak of serious disease in poultry. Eighteen people also became infected with the virus and six of them died. In order to bring the outbreak to an end all poultry in Hong Kong were slaughtered.

H5N1 viruses appeared in a number of Asian countries in 2003, in Europe in 2005 and in Africa in 2006. Millions of ducks, chickens and turkeys died from the disease or were slaughtered; H5N1 viruses also infected humans, causing severe respiratory disease, culminating in death in many cases. At about the same time H9N2 viruses emerged in Asia, causing disease in poultry with some transmission to humans.

Most, if not all, of the human cases of 'bird flu' caused by H5N1 and H9N2 viruses were in people who work with poultry and who presumably became infected as a result of direct contact with virus on the birds themselves, or in their faeces. Like many of the animal coronaviruses that have infected humans, these avian influenza viruses appeared to have little or no propensity for human-to-human transmission. H5N1 viruses have also infected other mammalian species including domestic cats; tigers and leopards at a zoo in Thailand died as result of infections that they acquired from eating virus-contaminated chickens.

Recently Discovered Viruses

Human metapneumovirus

In 2001 a new member of the family *Paramyxoviridae* was isolated from young children in the Netherlands. The virus was named human metapneumovirus. Infection can cause mild respiratory problems, but can also cause bronchiolitis and pneumonia, with symptoms similar to those caused by infection with respiratory syncytial virus, another member of the family *Paramyxoviridae*. There is serological evidence that the metapneumovirus had been present in humans for at least 50 years, and that in the Netherlands most children have become infected with the virus by the age of five. Infants and children with acute respiratory tract infection due to human metapneumovirus have since been reported in a number of countries, including the UK.

Re-emerging Viruses

Measles and mumps viruses

Further members of the family *Paramyxoviridae* are measles and mumps viruses. Both are important human pathogens, especially measles, which is a major cause of mortality in developing countries.

Until the late 1990s cases of measles and mumps had been declining in the UK as a result of widespread uptake of the measles, mumps and rubella (MMR) vaccine, but after fears were raised surrounding the safety of the vaccine fewer children were vaccinated. Measles, and especially mumps, began to 're-emerge', with hundreds of measles cases and thousands of mumps cases each year.

Bioterrorism

There are concerns that the world may witness the reemergence of a virus as a result of a deliberate release by terrorists. There is a long list of potential agents, including viruses, that terrorists might consider using. Smallpox virus, a member of the family *Poxviridae*, is high on the list. There have been no smallpox cases since the 1970s, smallpox vaccination programmes have been discontinued and most humans lack immunity to this virus. Some governments have stockpiled vaccine to be used in the event of a smallpox release.

Virus Surveillance

In order that the threats posed by emerging viruses and other agents can be dealt with effectively, it is important that world-wide surveillance systems are in place. There need to be effective systems that warn when agents such as the SARS coronavirus emerge.

The scheme for monitoring influenza virus strains provides a good model for a surveillance system. In this scheme, which is co-ordinated by the United Nations World Health Organisation (WHO), isolates of influenza virus from laboratories around the world are sent to Collaborating Centres for Influenza Reference and Research in London, Atlanta, Melbourne and Tokyo. Antigenic variation of the virus in humans and in animals (especially birds and pigs) is monitored, and twice a year the WHO recommends the virus strains to be mass-produced for incorporation into influenza vaccines.

Dealing with Outbreaks

It is important that resources for diagnosis, research and treatment of infectious diseases are available when the need arises. A useful case study is provided by the measures taken to deal with the emergence of the SARS coronavirus. There was some delay in alerting the world to SARS but, once it was apparent that this virus posed a major threat, work got under way in a number of virology laboratories. The virus was isolated in February 2003 and three months later its genome had been sequenced. The following year a paper was published reporting compounds that inhibit replication of the virus, while other papers reported the cell receptor of the virus and the structure of its replicase protein.

Diagnostic laboratory methods to detect evidence of SARS coronavirus in samples from patients rapidly became available; tests based on immunofluorescence and RT-PCR were developed. The health of international travellers was monitored and those found to be infected with the virus were nursed in isolation. These measures brought the SARS outbreak under control.

When there is an outbreak of a highly infectious virus, such as the SARS coronavirus or Ebola virus, infected patients and their contacts are quarantined. Control measures for some outbreaks include slaughter of animals that are infected and those that have been in contact with infected animals. This has been a key measure in dealing with outbreaks of foot and mouth disease (cattle and sheep), Nipah virus (pigs) and avian influenza (poultry). Medical staff, veterinary staff and other personnel dealing with humans and animals infected with high-risk viruses must take precautions to protect themselves and to avoid spreading infection. Precautions include wearing protective clothing and breathing filtered air.

INDEX